Summary

This guide presents a new way of working in construction. It provides an approach to enable all the key construction participants to agree ways in which their respective contributions can be maximised, while simultaneously achieving their own business objectives. In summary, each trades the benefits they can offer for fair reward.

The guide concentrates on the key principles of benefit trading and how to apply them. Part 1 of the guide defines benefit trading, introduces the fundamental principles and identifies the key steps involved. It shows how benefit trading is complementary to, though distinct from, other improvement initiatives. Part 2 examines the opportunities available for benefit trading in construction. Important tradable issues are identified for each of the main players. This part shows how benefit trading may be applied in a range of construction settings. Part 3 shows how to use benefit trading effectively. It describes what is involved in each of the key steps. It shows how trading is conducted and managed, how benefits are measured and how the process can be reviewed to ensure that valuable lessons are learned. The guide is designed to be used in conjunction with the accompanying tutorial and video, Publication C527V *Benefit trading – a practical guide for construction*

Benefit trading – a practical guide for construction

Potter, M and Connaughton, J

Construction Industry Research and Information Association

Publication C526

ISBN 0 86017 526 X

Keywords Benefit trading, integrated teams, collaboration, procurement, value.	
Reader interest Individuals and organisations using and providing construction services	**Classification** AVAILABILITY – Unrestricted CONTENT – Guidance STATUS – Committee-guided USER – Construction industry supply chain

Published by CIRIA, 6 Storey's Gate, Westminster, London SW1P 3AU.

ACKNOWLEDGEMENTS

This benefit trading guide and tutorial were written by:
Malcolm Potter
John Connaughton
of Davis Langdon Consultancy

and published by:
CIRIA (Construction Industry Research and Information Association).

The accompanying video was written and produced by:
Julie Wilkinson
of T.E.N (Television Education Network)

The project was funded by the Department of Environment, Transport and the Regions with contributions in kind from CIRIA, T.E.N and Davis Langdon Consultancy.

The preparation of the guide and tutorial was reviewed by a steering group comprising:

Dr Colin Robinson (Chairman)	KPMG
Mr Richard Aird	Richard Aird Associates
Mr Gordon Bateman	Thames Water
Mr Simon Bullivant	Roger Bullivant Ltd
Mr Colin Davidson	AMEC Construction Ltd
Mr Trevor Hursthouse	Goodmarriott and Hursthouse
Mr David Meek	Railtrack plc
Mr Andrew Bacon	Laing London Ltd
Professor Geoffrey Trimble	Loughborough University
Mr Tony Trinick	TriTone Partnership
Mr Mike Tweedie	WS Atkins Consultants Ltd

CIRIA's research manager for the project was Robert Dent.

The guide is designed to be used in conjunction with *Benefit trading in construction* tutorial and video (CIRIA publication C527V).

A large number of people and organisations, in addition to the foregoing, generously gave their time and experience to the project during the consultation phase. CIRIA and the authors gratefully acknowledge their contributions.

PREFACE

You should always expect a good deal from construction – whenever, as a customer, you buy from the industry or, as a supplier, you trade your services and products within it. Sadly, you don't always get what you expect.

Several recent initiatives seek to improve the performance and reliability of construction*. As a result, some fundamental changes are taking place that radically improve the way business is done, including:

- more collaborative working between all parties
- a new spirit of openness and trust
- the early involvement of key participants
- a commitment to improving design and construction processes and technologies, and reducing waste
- recognition that, to be effective and efficient, all parties need to give of their best and to receive fair reward and profit in return.

A crucial requirement is for all key participants – customers, designers, constructors, manufacturers and others – to work together more closely to develop better services, products and projects. To succeed they must communicate effectively and understand what each can bring to the construction process. More than that, they must agree ways in which their respective contributions can be maximised, while at the same time achieving their own business objectives. In essence, each must trade the benefits they can offer for fair reward.

Benefit trading is how we describe this new way of doing business. It goes well beyond traditional forms of negotiation which tend to focus on the settlement of accounts or resolution of disputes. Benefit trading concentrates on improving the value of construction for all who take part in it, both customers and suppliers. And it works equally well in the private and public sectors.

The essentials of benefit trading covered in this guide are simple but very exciting. Simple, because they are straightforward and accountable, and can be applied by customers and suppliers alike. And exciting, because they provide a beneficial and profitable basis for truly collaborative ways of working for all concerned.

* Principally, Sir Michael Latham's *Constructing the Team* (1994) and Sir John Egan's *Rethinking Construction* (1998).

THE PURPOSE OF THIS GUIDE

This guide will help you to:

- understand the fundamental principles of benefit trading
- identify the opportunities for benefit trading by learning from case study examples
- apply benefit trading in construction and in particular get the most out of collaborative working
- implement benefit trading as a part of an accountable trading process.

What the guide covers and how to use it

The guide concentrates on the key principles and how to apply them. It is accompanied by a Training Pack and Video that provide practical learning material to help you develop your own good practice in benefit trading.

The guide does not provide exhaustive guidance on every topic raised. References are given to sources of further information on key points as they arise in the text.

Who should read this guide

This guide is aimed at any individual or organisation using or providing construction services. It will be especially useful if you are embarking on collaborative working or considering doing so. You can use this guide whether you are a final customer or a product manufacturer, and regardless of your position in the construction supply chain.

PART 1
WHAT IS BENEFIT TRADING?

Part 1 defines benefit trading, introduces the fundamental principles and identifies the key steps involved. It shows how benefit trading is complementary to, but distinct from, other improvement initiatives in construction.

1.1 The essence of benefit trading 3

1.2 How it works – summary 4

1.3 Why use benefit trading 4

1.4 How benefit trading fits with other improvement initiatives 5

1.5 What makes benefit trading distinctive 6

PART 2
OPPORTUNITIES IN CONSTRUCTION

Part 2 examines the opportunities available for benefit trading in construction. Important tradable issues are identified for each of the main players. This part shows how benefit trading may be applied in a range of construction scenarios.

2.1 What can be traded? 9
2.1.1 Overview of tradable issues in construction 9
2.1.2 Price, skill, performance, methods 9
2.1.3 Improving production processes 11
2.1.4 Integrating design and construction 13
2.2 When can benefits be traded, and by whom? 15
2.2.1 Introduction 15
2.2.2 The key construction activities and phases 15
2.2.3 The key construction players 15
2.3 Benefit trading and construction procurement 18
2.3.1 Does size or position in the supply chain matter? 18
2.3.2 Does the procurement approach matter? 18
2.3.3 Application to common procurement approaches 22

PART 3
USING BENEFIT TRADING

Part 3 shows how to use benefit trading effectively. It describes what is involved in each of the key steps. It shows how trading is conducted and managed, how benefits are measured and how the process can be reviewed to ensure that valuable lessons are learned.

3.1 Introduction 24
3.1.1 The key steps in benefit trading 24
3.2 Step 1 – Developing the case for trading 26
3.2.1 The trading context 26
3.2.2 Identifying business needs 26
3.2.3 Identifying trading opportunities and trading partners 27
3.3 Step 2 – Committing to trading 29
3.3.1 Recognising the barriers to trading 29
3.3.2 How buyers can overcome them 29
3.3.3 How sellers can overcome them 31
3.3.4 Confirming the decision to trade 31
3.3.5 Planning and preparation 31
3.4 Step 3 – Exploring opportunities 32
3.4.1 First moves 32
3.4.2 Getting to know each other's business and processes/exploring options 34
3.4.3 Evaluating potential benefits 35
3.4.4 Developing the approach 35
3.5 Step 4 – Exchanging concessions 36
3.5.1 Getting started 36
3.5.2 Presenting the issues 38
3.5.3 Proposing and selling 38
3.5.4 Concessions bargaining 39
3.5.5 Making the agreement 40
3.5.6 The people involved 40
3.6 Step 5 – Measuring and reviewing the benefits 41
3.6.1 Benchmarking and mapping 41
3.6.2 Setting up benchmarks 44
3.6.3 Feedback and review of benefits 44

THE TOOLBOX

The reader's attention is drawn to the training material and checklists in the Toolbox and suggestions for further reading that may be used to amplify some of the points raised here and provide specific guidance on negotiating techniques.

T1 Procurement 49
T1.1 Lessons from other industries 49
T1.2 Procurement descriptions 50
T2 Trading tools 53
T2.1 Developing the trading plan 53
T2.2 Trading goals – proforma 55
T2.3 Assessing the trading opportunities – proforma 56
T2.4 Preparing your responses – proforma 57
T2.5 Trading pointers 58
T3 Measuring 61
T3.1 Benchmarking 61
T4 Policy and legislation 65
T.4.1 EC Regulations 65
T4.2 Delivering accountability 66
T5 Reading list and useful addresses 68

TRAINING MODULES
(provided in C527)

TM Instructions for course facilitators
TM.0 An overview of benefit trading
TM.1 The case for benefit trading
TM.2 Committing to benefit trading
TM.3 Exploring trading opportunities
TM.4 Exchanging concessions
TM.5 Reviewing/measuring benefits

CASE STUDY

Superstore is benefit trading superstar

A project completed 39 per cent faster and for 5.75 per cent less than a similar project built just 18 months earlier – impossible? No, just the result of excellent teamwork, co-ordination and ***benefit trading*** *from everyone involved – client, designer, main contractor and specialists.*

"We were looking for a quantum leap in terms of construction process" says the construction director. "The aim was to bring down construction time without compromising safety or increasing costs." The target was to build the store in 20 weeks and reduce costs by 10 per cent.

Before starting on site, the contractor called together the team of specialists to talk about how he wanted the team to work. This, the contractor explained, required everyone to challenge the obvious, strive for improvements, co-operate closely and trade mutual benefits arising from their achievement – without compromising their own profitability.

The contractor quickly got to know the client's requirements inside out. By communicating these to the specialist contractors, he enabled them to deliver even greater value. Together they achieved time reductions through off-site fabrication and modularisation, and also introduced design changes that improved the use of the building. For instance, omitting columns from the sales area not only made the building easier to construct but also increased flexibility for the client.

The 9250 m^2, £11.5 m project was completed in June 1999. Of course it was not without its problems and the client has planned workshops to review these and take forward the lessons learned to the next project.

These remarkable results were achieved by drawing on a combination of construction improvement initiatives including:

- **collaborative working** – team work involving openness and trust
- **early involvement of key specialist contractors**
- a critical **review of the design and construction processes**, involving the introduction of modularisation and off-site fabrication
- site **performance monitoring** and the application of feedback
- **benefit trading** – a positive working environment, promoted by the contractor, that:
  - encouraged key trades to integrate their work and engage in negotiating mutually beneficial solutions
  - acknowledged the importance of meeting the commercial goals of all parties
  - insisted on providing an efficient working environment for all, including the subcontractors.

This guide explains in more detail this exciting new process of **benefit trading** which will help to radically improve how the construction industry works.

1

WHAT IS BENEFIT TRADING?

Part 1 defines benefit trading, introduces the fundamental principles and identifies the key steps involved. It shows how benefit trading complements, but is distinct from, other improvement initiatives in construction.

'...costs are down and quality is up, we've been able to offer repeat business and even the specialist trade contractors have been able to invest in new products, and make a bigger profit. Everyone has been doing it – it's called benefit trading...'

1.1 THE ESSENCE OF BENEFIT TRADING

The essence of benefit trading is simple. It is a structured, value-based trading process that enables buyers and sellers to reach a deal that will benefit both of them. It works by focusing on the *benefits* that each party can offer the other including, among other things, the prices to be agreed.

There are five key principles. Those trading must be willing and able to:

- identify potential trading partners and issues of real benefit to them (tradable issues)
- understand and value tradable issues in both buyers' and sellers' terms
- trade issues on the basis of exchanging concessions – "if you will do this I may be prepared to do that…."
- ensure that both buyers and sellers stand to gain, but not necessarily equally. While trading may be tough, it should also be fair and based on trust – abusing trading power to extract concessions for free will, in the end, benefit neither party
- establish mechanisms for measuring the value of tradable issues and reviewing how well benefit trading is working.

Part 3 shows how the approach fits with the EC procurement Regulations, HM Treasury procurement guidance and the current emphasis on best value in local government.

In practice, benefit trading usually involves a number of different concessions:

- where each party can genuinely trade only one issue, the scope for benefit trading is reduced
- the scope is also reduced where the parties offer essentially similar concessions, although in practice similar issues are prioritised differently by each party.

Benefit trading can be used by the private and the public sector alike. Note that there is nothing in the current regulations governing the operation of public bodies and utilities that prohibits the use of benefit trading.

FIG 1 BENEFIT TRADING – THE KEY STEPS

1.2 HOW IT WORKS – SUMMARY

Part 3 provides further details.

Benefit trading involves five key steps (see Fig 1). While all the steps are important, on smaller, simpler projects some can be combined, for a more streamlined process. It is important to remember that the steps do not necessarily follow key stages in the construction process and benefits can be traded throughout the entire project life-cycle.

The key to benefit trading lies in the fact that buyers and sellers place different values on the issues being traded. Traders must therefore be knowledgeable about the issues on offer and be able to assess their value to others as well as to themselves. Without this knowledge they are unable to exchange benefits effectively. Thorough preparation is therefore critical to a successful outcome. Steps 1 and 2 involve setting up a trading relationship, including establishing a trading protocol, developing an understanding of the potential opportunities and overcoming any barriers to benefit trading.

Initial trading activity (Step 3) concentrates on exploring these possibilities and identifying and evaluating the options available for trading. Before concessions are exchanged (Step 4), it is important for traders to explore each other's businesses. Parts 2 and 3 of this guide show how effective trading is as much about knowing what is possible for the other party to do, as it is about knowing what value might be placed on what you have to offer.

Exchanging benefits (Step 4) is carried out on a concessionary basis – offers are responded to with counter-offers until each trader is satisfied with the proposed exchange and agrees the deal. Throughout the preparation and bargaining stages traders need to check and evaluate all the issues being traded against relevant benchmarks. This is a key part of the process – everyone involved in benefit trading must be clear about the benefits being achieved.

Part 3 provides detailed guidance on using benefit trading at each of these key stages.

The final step (5) produces a clear and accountable record of the process, including the benefits gained. Most of all, it reviews the whole process, identifies the valuable lessons learned and ensures that they are fed back to improve future performance at each key stage.

An essential precondition of successful benefit trading is willingness by both parties to trade in an open and collaborative manner.

1.3 WHY USE BENEFIT TRADING

Benefit trading is particularly relevant for construction because:

- there is great potential for the key players to add value at different stages during design and construction
- there are typically many issues to be traded between most of the key players, even on the simplest of projects
- all parties can benefit.

Improving the performance and reliability of construction depends on greater integration of design and construction processes. To do this effectively, the participants must be willing to work together and to contribute their best ideas and efforts. Benefit trading provides the means by which these contributions can be captured, valued and rewarded.

It encourages those who take part to sharpen their competitive edge by developing mutually advantageous and highly valuable practices and products. It helps them build long-term working relationships based on strong and profitable businesses – strong businesses that will form the basis for developing a healthy supply chain.

Why use benefit trading? Quite simply because it is good for business, both now and in the future. This guide will help you get the most out of benefit trading by adopting a structured approach to trading that focuses on the value-adding potential of the trading parties. This, not tactics that seek to exploit unequal trading power for the sole benefit of one party, is the key to improved profitability for all involved.

FIG 2 BENEFIT TRADING – THE KEY TO SUCCESSFUL CONSTRUCTION

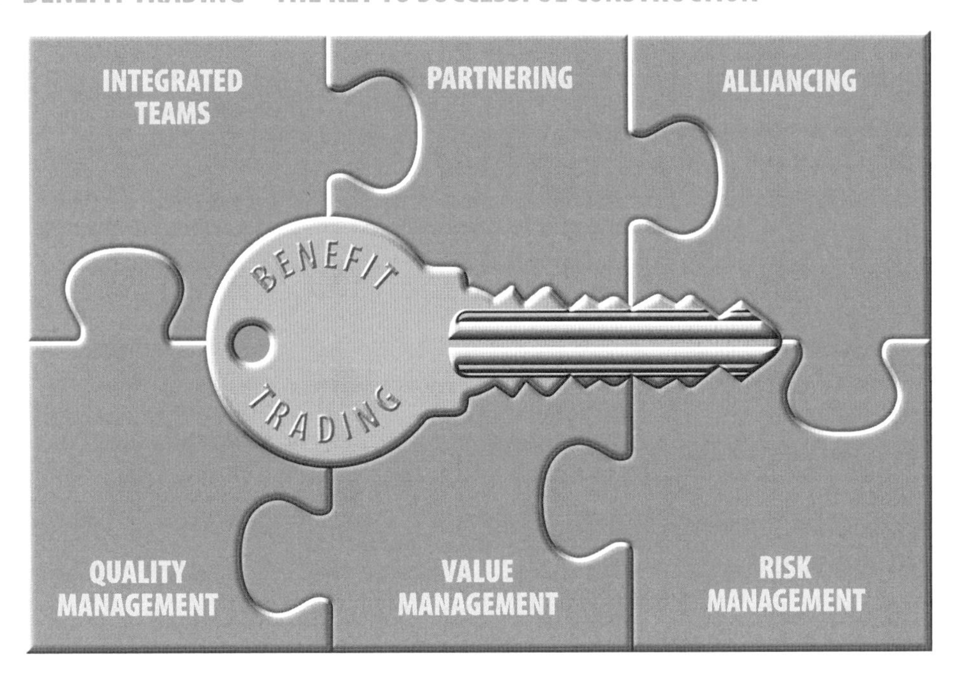

1.4 HOW BENEFIT TRADING FITS WITH OTHER CONSTRUCTION IMPROVEMENT INITIATIVES

Design and construction processes are being changed radically as many leading companies and their clients introduce new and improved ways of working. These methods include:

- integrated teams, including procurement arrangements that encourage the early involvement of key contributors to design and construction processes
- alliancing and collaborative working
- partnering
- a range of approaches that focus on meeting customer needs, such as:
  - initiatives for improving quality, including the introduction and development of total quality management
  - the development of innovative construction products and processes
  - risk management
  - value management.

But how is it possible to develop and foster the high level of co-operation between the key players that is needed for many of these approaches? More crucially, how can this co-operation be made to work commercially? Paying careful attention to team building, including the facilitation necessary to put collaborative working arrangements in place, can help answer the first question. Benefit trading is the answer to the second. It is an essential – and often missing – ingredient that is critical to the success of these initiatives because it allows the parties to:

- contribute their ideas, skills and resources for due reward and without fear of exploitation
- examine ways in which their activities relate to each other and, where possible, how they can be integrated and improved for mutual benefit
- identify key risks and, through a process of trading and negotiation, determine how best to manage them and on what terms
- work closely together and collaborate in genuine teamwork, developing positive working relations rather than adversarial ones.

Above all, benefit trading allows those with different interests to *improve their respective businesses*, especially those who have decided to work with each other to help define and achieve common goals. It offers the ultimate opportunity for win-win results.

1.5 WHAT MAKES BENEFIT TRADING DISTINCTIVE

Many of these improvement initiatives challenge traditional ways of working and require new tools and techniques to maximise their potential. Traditional approaches to negotiation, with their emphasis on price reduction as the ultimate goal, are not sufficient. Benefit trading provides a much more positive and mutually beneficial way of doing business, because it:

- encourages participants to look for and discover trading opportunities that may otherwise be overlooked, particularly upstream in the supply chain
- has a two-way focus – on the *seller* as well as the buyer – unlike more conventional approaches to negotiation
- is ultimately a commercial activity concentrating on the process of trading and reaching an agreement for genuine business benefit.

Parts 2 and 3 give guidance on the range of applications in construction.

Not only does benefit trading support many of these improvement initiatives, it is also an approach to developing business relationships and to improving design and construction processes in its own right.

2

OPPORTUNITIES IN CONSTRUCTION

Part 2 examines the opportunities available for benefit trading in construction. Important tradable issues are identified for each of the main players. This part shows how benefit trading may be applied in a range of construction scenarios.

FIG 3 THE BENEFIT TRADING OPPORTUNITY WELL

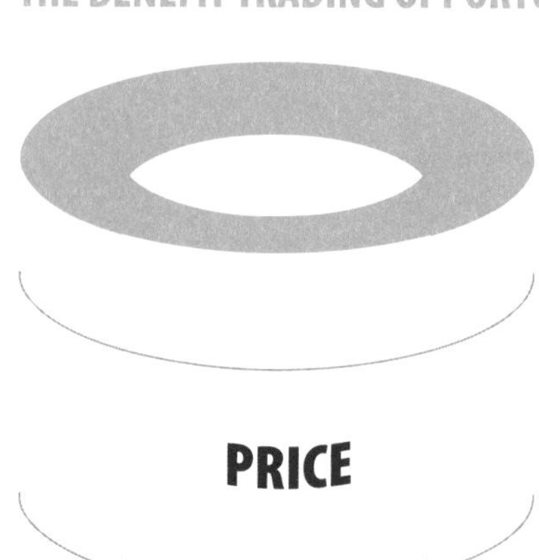

2.1 WHAT CAN BE TRADED?

2.1.1 Overview of tradable issues in construction

Benefit trading is not yet widely used in construction. Some current practices – for example, competitive tendering on price alone and the engagement of contractors at a late stage in the design process – are not, whatever their merits, generally conducive to benefit trading nor to achieving best value. While the former practice tends not to encourage the exchange and trading of information and ideas, the latter severely limits such an exchange.

A companion guide, *Selecting contractors by value,* was published by CIRIA in 1998 and shows how contractors can be engaged early in the process and selected by criteria other than price.

When trading does take place in construction – usually involving "negotiation" – it tends to concentrate on few well-known issues, such as price. The problem is that there are many other, more valuable issues that can be traded, but participants do not always have the will, the opportunity or the knowledge to seek them out.

Figure 3, the benefit trading opportunity well, identifies some of the more important tradable issues in construction. Those deep down in the well are potentially very valuable, but the parties that can benefit from them must be aware of this potential. And of course their owners must be willing to trade them. Benefit trading works best if both parties agree to seek improvement by uncovering these deep-lying opportunities. To do this effectively they must be willing to disclose some fundamental details of their own operations and be receptive to suggestions for change and improvement. The case study examples on the following pages show how companies have done this.

2.1.2 Price, skill, performance, methods

Part 3 provides guidance on priority setting and on clarifying trading objectives.

The first four topics in the well – price, skill, performance and method/buildability – are probably the most commonly traded issues. They are all interrelated: the price, for example, will depend on the particular combination of skill, technology, resources and methodology. Those trading will need to be able to explore different combinations of these elements to arrive at a satisfactory outcome. Also, particular issues, such as time, can be reflected in prices quoted and can also be tradable issues in their own right. Those trading need to be clear about their priorities so that they can, where appropriate, trade-off some issues against others. It is not always possible, for example, to get the most effective method at the lowest cost in the shortest time!

See also under 2.3.2 *Value Management* below

The potential for designers to add value through the creative application of available construction technology and specification is considerable. It is important that designers have the freedom to identify and offer alternative approaches that meet or exceed the client's requirements, and are able to trade – and thereby share in – the improvement benefits. Similarly clients may also want to trade-off elements of design or specification to improve value.

Process mapping of site activities can identify considerable potential for improvement in the efficiency of production and assembly operations, and site logistics. A great many site inefficiencies arise from the difficulties of co-ordinating different trades. Sequencing and other changes may need to be negotiated between a range of parties, all of whom may have significant benefits to trade for the overall good of the project.

Some negotiators tend to focus on reducing each others' margins as a way of reducing costs. This is misguided. Profit is normally only a small proportion of the price paid, and far greater savings are possible by targeting inefficiencies in production and supply processes, or by seeking alternative processes and products. In benefit trading it is important to recognise that both parties to a deal deserve fair reward – in the form of

CASE STUDY

The steel fabricator's story

A small-scale steel fabricator had a number of clients for whom it produced one-off assemblies. One client in particular, who had a regular building programme, was responsible for more than 30 per cent of the company's orders. This client proposed entering into a three-year supply arrangement with the fabricator on condition that it reduced prices by 5 per cent per annum and guaranteed certain delivery timescales. The fabricator welcomed the proposal but needed to look for ways to reduce material costs in order to achieve the proposed price reductions.

Until this time, the fabricator's turnover and cashflow had not been large or regular enough to persuade the steel stockists to provide credit facilities. As a consequence, it was obliged to purchase materials in stock at the time they were needed. Steel stockists seldom carry a full range of lengths of all standard sections, which means that buyers often have to purchase lengths considerably in excess of their needs and waste the surplus material when it is cut to size. This can amount to as much as 10–15 per cent of the value of the job.

As a result of discussions and subsequent trading between the client's buyer and the fabricator, they came up with a mutually satisfactory deal. In return for the fabricator meeting the price and delivery requirements, the client agreed to:

- provide the steel stockist with financial backing to enable the fabricator to be offered a credit facility
- standardise the assembly designs to facilitate the use of a defined range of standard lengths and sizes
- give three months' notice of supply requirements, thus enabling the fabricator to obtain a discount on bulk purchase standard supplies.

Lessons:

- both parties had to discuss and identify each other's needs before they could come up with a solution
- the concessions offered by the client cost little or nothing but were of considerable value to the fabricator
- by taking advantage of existing market opportunities, the fabricator was able to improve the company's margins and reduce costs to his client.

profit – for the benefits they offer. By doing this, parties not only create goodwill and a mutually supportive trading environment; they also ensure that their efforts are focused on where the value-adding potential is greatest, not on reducing profit.

While price, skill, performance and method are the issues most commonly traded in construction, they are often poorly understood. Take price, for example. Most estimators would claim to know the 'going rate' for building materials or work used regularly. But how many know enough about the production process to enable them to negotiate efficiency cost savings while still maintaining the required quality?

2.1.3 Improving production processes

The next set of topics lying deeper down the well is primarily concerned with eliminating waste and improving efficiency in supply and production processes. Manufacturing industries have been successfully focussing for some time on the reduction and elimination of waste throughout their supply chains. The key lesson for construction is that this can only be achieved if those trading have a thorough knowledge of all relevant production processes.

The range of industries providing materials and components to construction is enormous – it would not be practical for this guide to review the great variety of opportunities for benefit trading that exist. Rather, it is important to understand how to search for real value-adding benefits and to identify potentially tradable issues that may exist in products and processes. The case studies on the next page illustrate the potential that can be unlocked by understanding the relevant production and supply processes.

Valuable insights into, and an understanding of, production and supply processes can be gained by:

- mapping the relevant processes, both on-site and off-site, including supply and delivery processes
- holding regular discussions with specialist suppliers and contractors about the potential for improvement and reduction/elimination of waste
- visiting suppliers' factories and other places of production to see at first hand the key elements of production and how the process is organised
- agreeing to share ideas and innovation through construction networks and demonstration projects.

WASTE

Reducing, reclaiming and recycling waste provide plenty of opportunities for benefit trading

CASE STUDY

Cutting out waste

A projects director for a major contractor was paying a regular visit to site and noticed two skips full of plasterboard offcuts. In response to questioning, the site agent advised that the skips had to be removed twice a day because of the quantity of waste being generated.

Returning to head office, the projects director phoned the production director at the plasterboard manufacturer's head office to enquire whether it would be possible to obtain boards cut to length.

The two companies were already in the process of establishing a trading alliance and the production director agreed to look into the matter and report to the next alliancing workshop, due to be held the following week.

As a part of a package of issues, a deal was struck a few weeks later that enabled the contractor to purchase non-standard lengths of plasterboard for any order exceeding 250 m^2 for the same price as standard sheet sizes.

CASE STUDY

Getting to know your suppliers

A new managing director had recently been appointed to a medium-sized contractor. The MD was well known for a hands-on approach and was expected to bring changes to the company. The board's reaction was distinctly unenthusiastic as one of the first initiatives was announced: "...from now on every director will be expected to visit at least three key suppliers every month". Reluctantly they all agreed, privately thinking it would be a waste of their valuable time.

At a board meeting six months later they reviewed the success of the "visiting suppliers initiative". Much to their surprise they had unearthed some very valuable information:

- the core business of some suppliers was not what they had expected it to be. For some years they had been inviting tenders for goods that some suppliers no longer produced, and in two cases never had
- they had discovered some interesting products and techniques of which they were previously unaware
- they had begun to learn about some of their suppliers' key business issues and started to talk about how these could be addressed.
- they had established working relationships with key personnel.

On the strength of their experience, the board agreed unanimously – and enthusiastically – to continue the practice and encourage the buyers and project managers to do the same.

Process mapping of off-site and delivery processes can be done with the help of the specialist supplier/contractor, and regular discussions will help identify potential for savings and improvements. Buyers who visit their main suppliers' factories regularly are able to keep up to date with current production methods and gain a better understanding of how these can be used to best advantage for their own requirements. They become better informed and they also have the opportunity to question the relevance to them of particular practices and processes. This is not a recipe for meddling; if they understand how the product is produced and delivered, buyers can indicate more clearly to the supplier where the real value lies. They can also suggest approaches that would, from their perspective, improve the product or process.

Regular discussions with suppliers and others in the supply chain not only provide the opportunity to explore important business issues and find solutions to mutual problems. They also help build good working relations that can encourage a more proactive and collaborative approach to getting the job done.

See Preface and Part 1.4.

2.1.4 Integrating design and construction

The ultimate goal of many of the improvement initiatives reviewed earlier is a more integrated design and construction process. By reaching deeply into the opportunity well, those involved in design and construction can come closer together and develop more integrated products and processes. Indeed, in a truly integrated design and construction process there would be fewer new opportunities for benefit trading as the key parties would already be working together and trading openly in a spirit of co-operation.

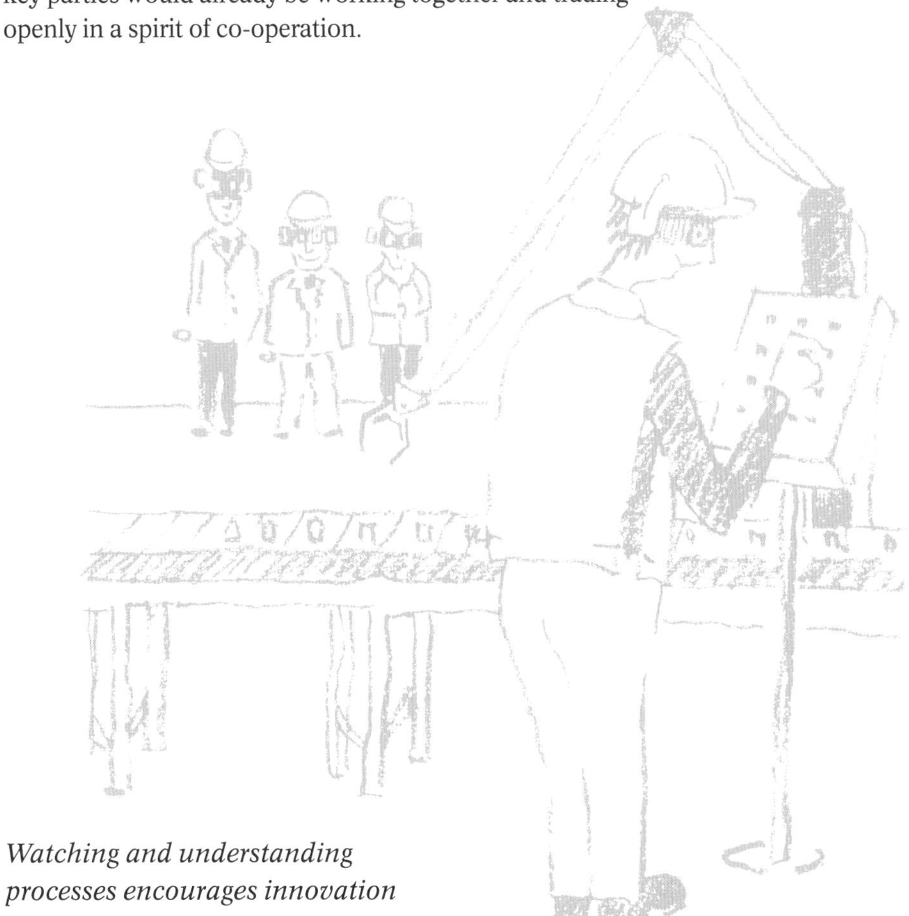

Watching and understanding processes encourages innovation

CASE STUDY

Integrating construction

A ten-storey office development had reached the fitting-out stage. The foreman responsible for dry lining had been called to the top floor to sort out a problem. The immediate problem was swiftly resolved but the search for a solution had led the foreman to examine the arrangement of the secondary steelwork to the lift shaft and question whether this could be better integrated with the dry lining.

The dry lining foreman took the opportunity to raise the concern over the lift shaft structure at the next trade contractors' review meeting. The lift manufacturer agreed to discuss the issue in detail and report back to the next meeting. Both trade contractors had a mutual interest in this exercise as they had other projects in the pipeline with the same contractor.

A detailed review of the dry lining and lift shaft configuration revealed that making a minor alteration to the secondary lift structure would enable the dry lining subcontractor to make use of it and omit a number of supports for his partitioning.

The two trade contractors were able to report back positively to the next review meeting. They had been successful in removing a small quantity of redundant structure and, in return for this saving, the dry lining subcontractor had agreed to fit the lift engineer's door linings while erecting the surrounding partitioning.

2.2 WHEN CAN BENEFITS BE TRADED AND BY WHOM?

2.2.1 Introduction

In this section we develop and expand the opportunity well of tradable issues outlined in Part 1 to take account of:

- the time they arise in the process – opportunities occur at different times as a project develops through design and construction, to completion and use
- the parties who can identify and trade them – different opportunities arise for the different people involved, because individual players typically undertake a range of activities at different times in the process.

Issues can be categorised to help identify tradable opportunities at different stages in the construction process, and for the key players involved as follows:

Method	how the work will be done
Resources	the intellectual, physical and financial resources available to do the work
Terms	the terms of the agreement to do the work, eg price and payment terms, terms of contract, risk allocation, etc

2.2.2 The key construction activities and phases

The following construction activities/phases provide a useful framework with which to review traditional opportunities that are available to the key players (note that they will not always occur in this particular form or sequence):

- feasibility, option appraisal and strategy formulation
- scope definition, design development and specification
- construction, including scheduling/programming and the development of the process itself
- supply logistics and delivery arrangements
- commissioning and handover
- use and maintenance.

Table 1 provides a checklist of issues for each of these activities/phases of work that can be traded by the key players.

2.2.3 The key construction players

The key players typically include:

- clients
- design team members
- general contractors (including design and build and management contractors, and construction managers)
- specialist trade contractors
- specialist suppliers.

TABLE 1 EXAMPLES OF GENERIC TRADABLE ISSUES FOR KEY ACTIVITIES

ACTIVITY/PHASE	METHOD How the work will be done	RESOURCES The intellectual, physical and financial resources available to do the work	TERMS The terms of agreement to do the work
STRATEGIC ADVICE • feasibility • option appraisal • strategy formulation	Work plan Communication plan Research Surveys Options to be appraised Scope of feasibility studies	Team and people Skills mix – strategic analysis – cost/financial advice – legal advice Funding sources Specification of equipment/facilities: – surveying/testing – hardware/software – communications systems Access to data – survey – costs/funding	Conditions of appointment Duties/responsibilities Output requirements Services required Indemnity insurance Payment terms Targets – time – cost – quality
DESIGN • scope definition • design development • specification	Design approach/philosophy Design management plan – information requirements – information sequence – information structure Communications plan	Team and people Skills mix – project management – design/buildability – cost advice and management – value engineering – risk management – quality management – health and safety – energy management Specification content – functions and spaces – material / components – servicing strategy Equipment/facilities – (as for Strategic advice) – accommodation Access to data – briefing/survey specification/product – cost	Conditions of appointment Duties/responsibilities Boundaries/interfaces with other roles Compliance with regulations Output requirements Services required Indemnity insurance Payment terms Targets – time – cost – quality Risk sharing/reward
CONSTRUCTION • construction process development • scheduling/programming • assembly and installation	Construction plan Phasing/sequencing Work packaging Temporary works/arrangements Work interfaces Access Material storage/handling Waste – reduction – removal – recycling Protection Security Health and safety plan/file	Team and people Amount of resources Skills mix – project management – buildability advice – construction/production control – cost information/management – value engineering – risk management – quality management/control – health and safety – specialist advice Cashflow/credit arrangements Materials/components – temporary – permanent – waste use/reduction/removal Equipment/facilities – accommodation/catering – storage – off-site/on-site plant – communications Data – drawings/specification/product – cost	Contract conditions Duties/responsibilities Boundaries/interfaces Compliance with regulations Scope of works Services offered Insurance Payment terms Rates – labour – materials – plant – preliminaries – overheads Profit Targets – time – cost – quality – warranties/bonds Risk sharing reward Liability for defects

TABLE 1 CONTINUED

	METHOD How the work will be done	RESOURCES The intellectual, physical and financial resources available to do the work	TERMS The terms of agreement to do the work
SUPPLY • supply logistics • delivery arrangements	Supply logistics – planning/sequencing Scheduling Ordering Checking Invoicing Health and safety plan/file	Team and people Manufacturing processes – pre-assembly – standardisation Skills mix Supply logistics Access to supply networks Cashflow/credit facilities Facilities/equipment – transport – storage – material handling Access to data – suppliers/manufacturers – costs	Supply conditions Outputs Insurance Rates – unit cost – transport/handling – storage – fixing Conditions and methods of payment – order sizes – discounts – frequency Targets – order times – availability – delivery – frequency/punctuality – price – quality Guarantees Maintenance Risk sharing reward
COMMISSIONING / HANDOVER	Commissioning plan Handover/occupation plan Defects management Training plan for users/facilities managers Health and safety plan/file	Team and people Skills mix – equipment specialists – facility management Equipment – controls – fittings – communications/data distribution Information/knowledge – manuals/instructions	Contract conditions Performance liability Insurance Conditions and methods of payment Rates of payment Targets – achieving design standards – compliance with regulations – time – price Guarantees/certificates
MAINTENANCE	Maintenance plan Facilities management plan	Health and safety plan/file Team and people Skills mix Funding Equipment – maintenance materials – replacement parts – access Data – maintenance manuals – costs – access to replacements	Maintenance contracts Working conditions Insurance Conditions and methods of payment Rates of payment Targets – specification – frequency – timeliness – responsiveness Guarantees

In practice, of course, many of these players have their own networks of suppliers and specialist trade contractors but, for ease of illustration, only these key roles are featured in this section.

Table 2 identifies the key issues that these participants may typically identify for trading:

- as *buyers*, in terms of what they *want*
- as *sellers*, in terms of what they *want*.

With the exception of clients, who are not normally selling within the construction process, all the other players may have dual roles of buyer and seller. While many players have a number of *wants* in common – for example, the need to receive a reasonable reward – they also have wants particular to their situation. Clients, when buying, place a high premium on predictability, reliability and quality of work. Contractors, when selling, may place a similar priority on obtaining continuity of work, and in return may be prepared to pass on some of the efficiency benefits obtained from increased volumes of sales or repeat work.

Part 3 provides further guidance.

Tables 1 and 2 cannot take account of all the motivations that exist in practice, but they will help to build a profile of the key players' wants and offers in preparation for benefit trading. Start by identifying the issues in each phase in which you have an interest, and consider your role as either buyer or seller. What do you want, and what will you offer in return? What do you think other parties can offer, and what will they want from you?

Clearly not all issues can be anticipated before they arise, but identifying as many as possible in advance will provide a useful basis for making progress. It also has the advantage of raising questions that might otherwise remain unasked.

2.3 BENEFIT TRADING AND CONSTRUCTION PROCUREMENT

2.3.1 Does size or position in the supply chain matter?

Part 3 provides further guidance.

The principles in this guide can be applied by all those who wish either to buy, or buy and sell, within construction, regardless of their position in the supply chain. However, traditional attitudes in construction tend to reinforce a hierarchical pecking order that can place those further along the supply chain at a disadvantage.

It is important – particularly for downstream or smaller firms – to remember that buyers need sellers. Sellers who offer the potential to add value will always be in demand. It is essential for all sellers, regardless of their position in the supply chain, to value what they can offer in the other party's terms.

2.3.2 Does the procurement approach matter?

The need for a positive trading environment

Benefit trading can be used within a range of construction procurement approaches but it is most effective when the strategy positively encourages:

- collaborative working between participants who are appointed at the right time and on the right terms
- timely and relevant contributions from the key participants
- the development of the construction supply chain
- a willingness to understand the design and construction processes through close scrutiny, feedback and review

TABLE 2 EXAMPLES OF BUYERS' AND SELLERS' WANTS

ACTIVITY/PHASE	METHOD How the work will be done	RESOURCES The intellectual, physical and financial resources available to do the work	TERMS The terms of agreement to do the work
BUYERS' WANTS Buyers from the construction supply chain mostly want the same kind of things	**Means** Effective management Best practice methods Sustainable processes **Timing** Work sequence and timing to suit business needs **Safety** Safe construction practice	**People** Good team – the right people – available – flexible **Skill/experience** Skilled workforce – experienced and expert – understanding the businesses' needs – access to specialist expertise **Funding** Financial stability **Facilities** Up-to-date – accommodation – equipment/plant – compatible communications **Supply** Access to – materials/components – supply/distribution networks **Information/knowledge** – cost – materials – regulations	**Performance liability** Service – clear service definition – single point responsibility Definition of roles Risk – responsibility for own work – low risk – indemnity against risk **Output requirements** Delivery – to defined standards – on time – within budget – free from defects – good value **Payment** – predictable price – payment for completed work – simple accounting **Guarantees** Warranted work
SELLERS' WANTS Sellers to the construction supply chain mostly want the same kind of things	**Means** Freedom to manage/produce efficiently Communication issues – quick clear decisions – no/minimum changes Frequent and simple payment/ invoicing **Timing** Timing to suit work sequences and production methods **Safety** Safe working	**People** – single point of authority – constructionwise Freedom to choose own – consultants/managers – labour – suppliers **Facilities** – compatible communications – unrestricted access to site – essential support facilities (storage, services, plant) **Funding** – funding certainty – investment for development – broad buyer base **Supply** – work to match skills – repeat orders – repetition – steady workflow **Information** – access to key personnel – access to essential data – statement of requirements – access to site/plant	**Performance liability** Clear definition – services outputs/requirements – responsibilities – roles **Risks** – control of own risks only – limit to guarantees and warranties **Conditions and methods of payment** – fair reward – to be profitable – payment related to type and quantity of work – prompt payment – regular payment **Targets** Achievable targets – time – cost – quality **Guarantees** Indemnity of own services only

Note: These are examples only. This is not intended to be an exhaustive list.

- value management and similar techniques for defining project objectives and delivering them in the most effective way
- risk management
- a clear focus on safety and quality and on meeting customer needs.

Collaborative working

Collaborative working cannot really be successful without benefit trading. Similarly, a genuine will to work together – whether involving the formation of alliances or the development of partnering relationships – provides a sound basis for effective benefit trading. It enhances participants' confidence in their position as team members and encourages the development of the trust required for ideas exchanged willingly. Building teams does not just happen, it requires effort and in many cases the use of facilitation to ensure that team relations are well rooted and able to develop effectively.

Specialists who may be reluctant to reveal commercial secrets in a competitive arena do so more willingly in a collaborative context. Their motivation for this is simple but important: once within a collaborative relationship they become committed to the team and are no longer fearful of being exploited by unscrupulous buyers. They soon realise that benefit trading is good for business. Toolbox T1.2 provides details of the more common forms of collaborative working in construction.

See also the companion CIRIA guide *Selecting contractors by value*.

Timely and relevant contributions from key participants

Procurement strategies that encourage key specialist trade contractors and suppliers to take part in design development increase the opportunity for more imaginative and technically robust solutions. Invariably, they also encourage a more efficient approach to design and to the development of production information by short-circuiting repetition and unproductive duplication.

Development of the construction supply chain

Procurement approaches that are highly prescriptive and inflexible provide little opportunity for dialogue or interactive decision-making. They are unlikely to offer much scope for benefit trading. Approaches that encourage dialogue and performance-based (rather than prescriptive) solutions are more likely to foster innovation and creativity and lead to improvements in service and value. Sellers can only really develop customer-focused products and services if they have regular access to, and dialogue with, those who specify, purchase and use their products and services.

Understanding design and construction processes

Improved understanding of design and construction is critical to the development of successful benefit trading. To derive real value from the approach, participants must be able to go deep into the opportunity well (Figure 3) and, in the process, increase knowledge and explore opportunities for future trading. Procurement strategies that include a commitment to regular review and continuous improvement help provide a systematic basis from which to challenge existing work methods, measure productivity and stimulate improved performance.

See also the companion CIRIA guide *Value management in construction: a client's guide*.

Value management

Benefit trading can add significantly to value management and value engineering (VM and VE). VM and VE offer structured and positive ways of capturing the creative potential of all project participants. They work by establishing a series of challenging interventions in the design and construction processes that test options and question key decisions. They encourage participants to work as a team to clarify objectives and then to achieve them in the most effective way.

FIG 4 PROCUREMENT APPROACHES AND OPPORTUNITIES FOR TRADING

MANAGEMENT

- A construction manager is appointed to supervise construction (and design)
- Construction services are appointed on a package basis
- Key construction providers are appointed early in the process
- Is intended for risky, complex projects

DESIGN & BUILD

- The contractor takes responsibility for design and construction*
- Once the design is agreed there is usually little room for change
- May use an in-house team or external consultants for design

* *The client may appoint own design team to define the project scope*

DESIGNER-LED LUMP SUM

- The design team provides all construction information for tender**
- The contractor has little or no involvement in design development
- Selection is usually based on price alone

** *In some cases expert advice may be sought prior to tender*

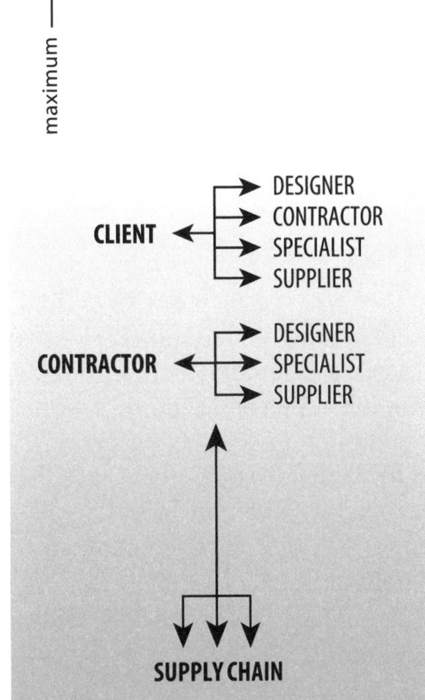

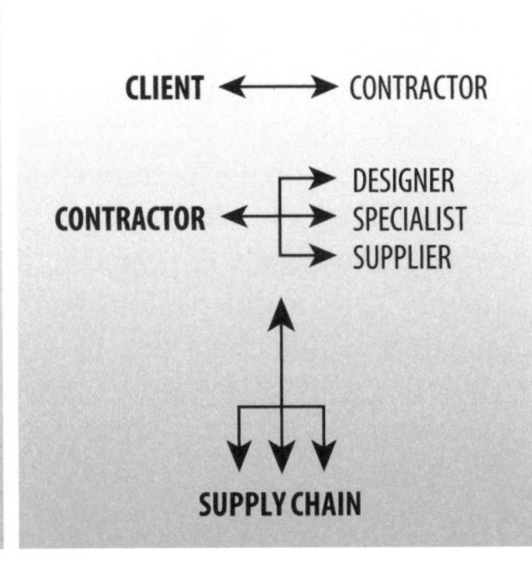

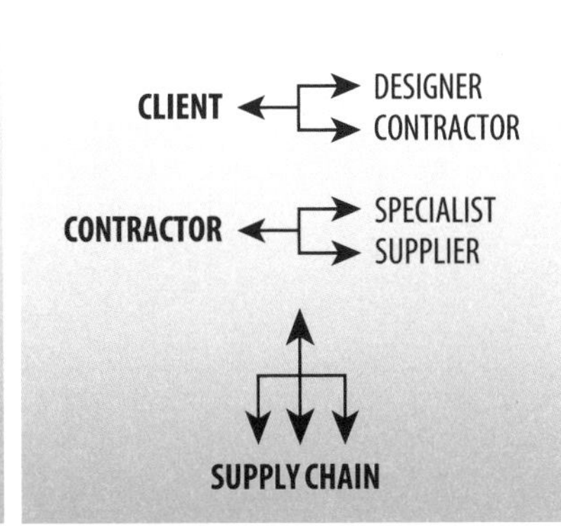

PRINCIPAL TRADING OPPORTUNITIES FOR KEY PLAYERS (TRADING INITIATORS IN BOLD)

OPPORTUNITIES

Management

- Client has significant trading opportunities for negotiations with all parties
- All parties may negotiate with each other

Design & Build

- Client has single opportunity for trading – with contractor
- Contractor may have many separate opportunities for negotiations with designers specialists and suppliers

Designer-led lump sum

- Generally, opportunities for trading may be limited if construction details are completed before tender
- Contractor may have many separate opportunities for negotiations with specialists and suppliers

This kind of working is ideally suited to benefit trading, as it requires interactive and constructive debate about the best means for achieving value for money. Benefit trading helps unlock the potential of the approach, by allowing participants to benefit directly from the improvements they identify and can deliver.

See also the companion CIRIA guide *Risk management in construction*.

Risk management

Risk management involves identifying major risks, assessing their likelihood and impact, and planning how best to deal with them. This process provides an important opportunity for benefit trading. Risk is one of the more tradable issues in construction, and one where the different perspectives of the key players – who may have different attitudes to risk and different opportunities to avoid or manage it – can be used to considerable benefit.

A focus on quality and customer needs

Benefit trading is explicitly value-based and customer-focused. It provides a means whereby real value can be identified and assessed, and will flourish in circumstances where customer needs, whether of initial buyer or ultimate end-user, can be clearly defined.

2.3.3 Application to common procurement approaches

The principal differences between the commonly used procurement strategies can be summarised as:

- the extent to which construction specialists are encouraged to contribute to the design and construction processes at an early stage
- the way that risk (and responsibility) is distributed between the parties
- the extent to which different activities in the design, manufacturing and assembly processes overlap one another.

See also the companion CIRIA guide *Planning to build?*

Figure 4 identifies the key opportunities for benefit trading under commonly used procurement approaches in construction. The comments are, inevitably, generalisations. There are many permutations to suit different needs in practice. However, the approaches that seek to overlap design and construction processes appear to have greater numbers of opportunities for benefit trading than those that don't. This is mainly because the relationships between the parties are more complex. It must be remembered that there can be a large number of separate contractual relationships in even the most simple construction project.

3

USING BENEFIT TRADING

Part 3 shows how to use benefit trading effectively.
It describes what is involved in each of the key steps.
It shows how trading is conducted and managed,
how benefits are measured and how the process can
be reviewed to ensure that valuable lessons are learned.

3.1 INTRODUCTION

3.1.1 The key steps in benefit trading

In Part 1, benefit trading was summarised in terms of five key steps shown in Figure 1. Following these steps will help you to get the most out of benefit trading.

That is, you should:

Step 1	Define your needs and identify your trading opportunities and partners
Step 2	Remove any existing barriers and confirm your decision to trade
Step 3	Get to know the relevant business processes and evaluate tradable issues
Step 4	Discuss possible concessions, bargain and agree a deal
Step 5	Review the process, measure benefits, benchmark performance, feedback results

FIG 1

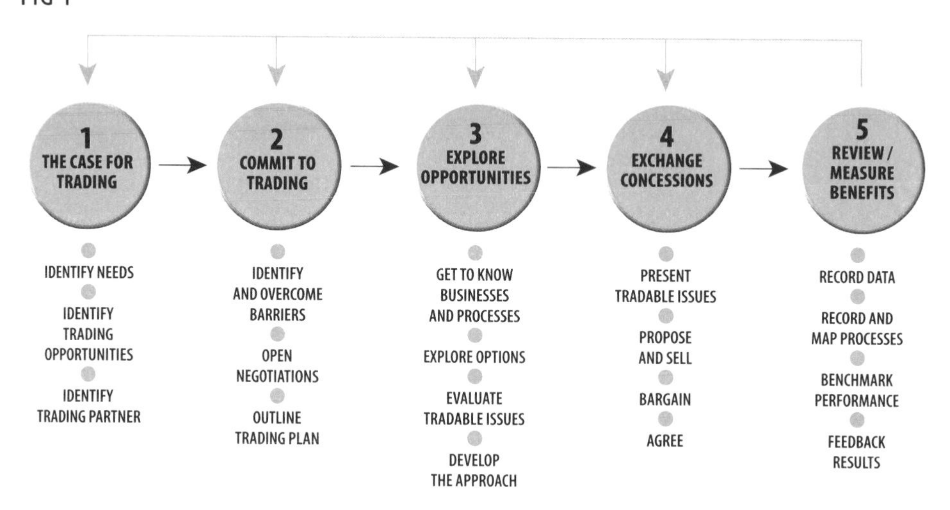

In practice, the steps are not rigid. On small, simple projects, for example, steps can be combined to streamline the process. For instance, Steps 1 and 2 can be brought together, as can Steps 3 and 4.

Also, while the steps follow a logical sequence, there will normally be some repetition/overlap between them. For example, new issues arising during trading may need to be evaluated against initial aims and objectives. Similarly, it is better to measure and benchmark benefits throughout the process rather than leaving it all until the very end.

Guidance on developing a trading plan is provided towards the end of Section 3.3 (Step 2 – Commiting to trading).

However, the overall sequence holds good for most situations. This section takes you through the benefit trading process step by step. Be prepared to modify the steps and the sequence to suit your needs. We recommend that you draw up a **trading plan** at an early stage in the process.

CASE STUDY

The contractor and the bricklaying subcontractor

This example of benefit trading shows that a good deal can be obtained from negotiation where there is an opportunity for interactive dialogue that can reveal information that would otherwise remain undisclosed.

A bricklaying subcontractor has offered a price for brickwork based on a specification that indicated:

- the quantity required
- the dimensions and profiles of the work
- the brickwork bond and quality requirements
- the specification of the bricks (to be supplied by others)
- the quality of pointing (to be laid overhand because of access difficulties and work sequencing)
- the start date
- the timescale for completion
- the method of payment (monthly).

The subcontractor has just had a job cancelled for which he had ordered a similar quantity of slightly better quality bricks at a time when the price was low. They are available and ready for delivery. He is faced with paying a 5 per cent cancellation charge if the order is not confirmed.

The subcontractor urgently needs the work, but only has one bricklayer, Fred, who is sufficiently skilled to undertake it. Fred will not be available until five weeks after the job is due to start. Taking on someone else will mean paying higher rates. The price submitted was keen but reflects the difficulties involved.

The contractor wants to use the subcontractor

- on account of a track record of good workmanship and reliability
- but the contractor needs to reduce the price of brickwork by about 10 per cent
- so the contractor decides to negotiate and calls the subcontractor in.

They trade concessions.

The subcontractor,

- who will have to pay 3 per cent more to hire another bricklayer, offers to reduce the price by 2 per cent
- if the contractor delays the start of the work by five weeks.

The contractor agrees

- since the start of the brickwork is not critical
- but only if the subcontractor is prepared to reduce construction time by two weeks.

The subcontractor

- offers to supply the bricks already ordered for the cancelled job, at a price significantly lower than the contractor is going to have to pay
- but only if the contractor will provide additional scaffolding to avoid overhand working.

The contractor

- agrees to accept this offer (as he will make a net saving of 7 per cent)
- but only if the subcontractor will provide labour to distribute the bricks around the site and guarantee their delivery
- the contractor's offer still leaves the subcontractor with a 2 per cent net gain.

Finally, the subcontractor,

- who normally pays the workmen weekly
- has allowed for 1.5 per cent bank charges in the tender to cover interest rates
- the subcontractor is therefore prepared to offer a further 1 per cent reduction
- if the contractor is prepared to pay for the work on a weekly basis
- the contractor agrees to this, provided that the subcontractor agrees that Fred is available to do all the difficult brickwork.

They agree the deal.

3.2 STEP 1 – DEVELOPING THE CASE FOR TRADING

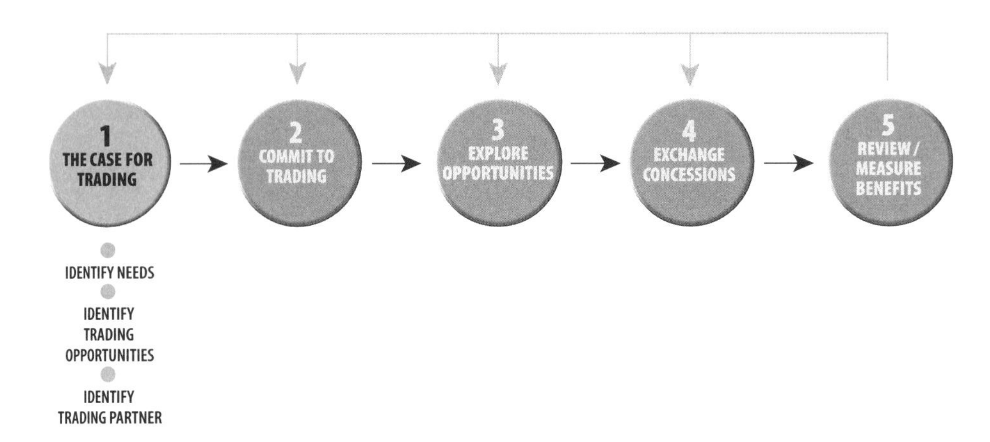

3.2.1 The trading context

In collaborative working

Benefit trading is essential to get the most out of collaborative working, whether as part of a strategic alliance, joint venture or partnering arrangement. Those already involved in such arrangements will know much about the other parties, their work and business drivers; in this case the need for exploratory work to identify tradable issues and partners is therefore reduced. Nevertheless all parties should be clear about their aims and objectives. Existing collaboration should not be seen as a reason for failing to identify and examine carefully all the potentially valuable tradable issues.

In traditional working relationships

There is considerable scope for using benefit trading where the parties have been brought together on a more traditional basis, eg as a result of competitive tendering. However, they will have to work hard to create an environment in which benefits can be identified and traded openly. This is because competitive tendering on price tends initially to limit the opportunity for contributions from the key parties. However, even in that situation, the value that the parties can add at different stages in the construction process, and the large number of tradable issues that are typically available in construction, mean that benefit trading can play a key role. This is shown in the case study opposite.

3.2.2 Identifying business needs

The first task is to define your overall aim and trading objectives. While your overall aim will be the position you would like to be in as a result of the trade, you will need to achieve several objectives along the way by exchanging concessions with the other party. These objectives can be traded for the concessions you offer, and you need to be able to identify what may be offered in return for them, and what the other party will want from you.

The starting point is to identify your objectives (ie the concessions you would like from the other party). You should define these in terms of their relative importance to you:

- *needs* – essential requirements
- *wants* – significant and useful though non-essential requirements
- *desires* – "nice to have" non-essential requirements.

Normally it is relatively straightforward to define your essential *needs*. This is the fixed point below which you are not prepared to trade. Your wants may be slightly more difficult to assess and you may find it necessary to adjust your expectations during trading. Your *desires* represent something of a bonus position beyond what you realistically expect to achieve; these are inevitably flexible and, in practice, are the focus of a good deal of trading.

Tables 1 and 2 in Part 2 identify some of the more important tradable issues in construction, and who owns them.

3.2.3 Identifying trading opportunities and trading partners

Once you have identified your aims and objectives, you should draw up a long list of all the possible items you might be willing to trade to obtain them, and with whom you could trade them. In practice the distinction between the concessions you want and those you may be prepared to offer is not so clear, but you should at least record them separately – this helps assess what is of real value to you and to the other parties.

See Annex T3 for information on process mapping.

You should be prepared to look at all the possibilities, even those which at first sight appear to be unpromising. This is really an information-gathering exercise (issues/concessions will be evaluated a little later in the process) and only issues that are clearly unacceptable to the other parties should be discarded at this stage. Techniques used for gathering information at this stage include:

- facilitated brainstorming
- structured interviews with key personnel
- review/analysis of previous projects and/or experience with the other parties, if any
- process mapping.

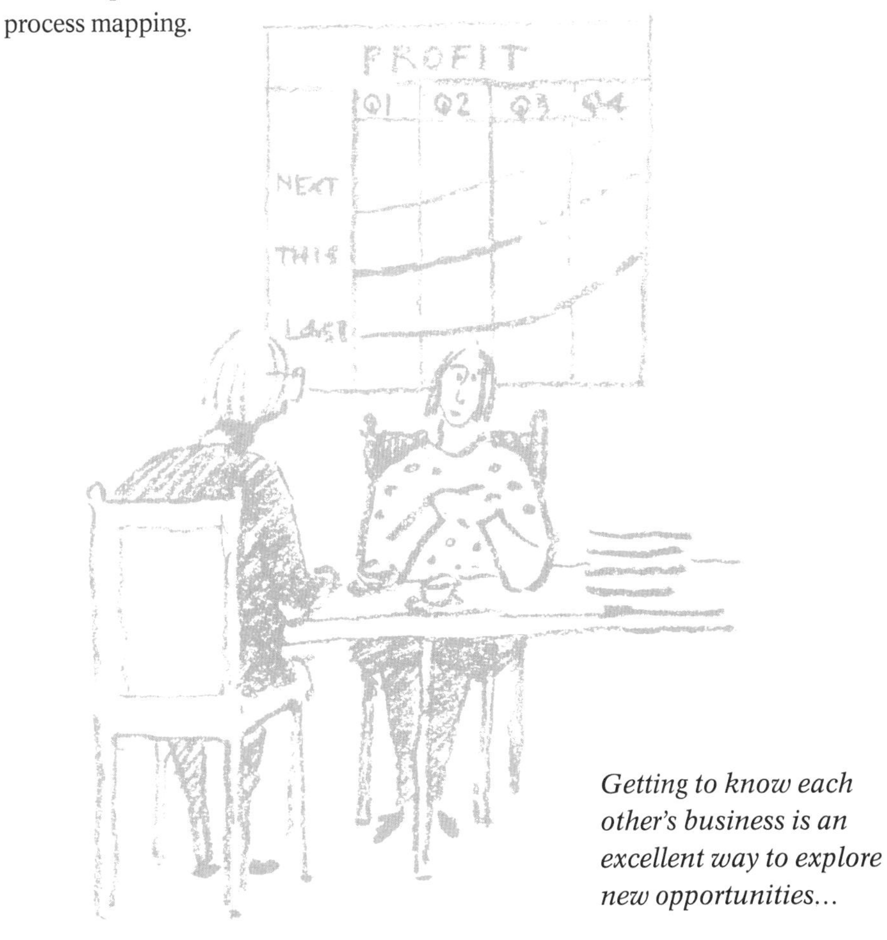

Getting to know each other's business is an excellent way to explore new opportunities...

CASE STUDY

How a local authority took up benefit trading

A county police authority decided to partner its £12 million construction programme with four regionally based contractors. The county authority's technical team, responsible for setting up and managing the arrangements, was keen to make the new venture a success.

The team's first action was to consult the internal auditor and discuss its proposals for complying with the European procurement procedures. The auditor was helpful, and between them they agreed a two-stage competitive approach:

Stage 1
Selection from a long list using a set of predefined management criteria.

Price was not an issue.

Stage 2
The twelve short-listed contractors priced a notional bill of quantities and specification based on three typical projects: a custody suite, an office block and a car park.

The final selection took account of information gathered from both stages and was made on the basis of weighted scores for price (60 per cent) and quality (40 per cent). The four most suitable contractors were appointed for a period of three years with a further negotiable two-year extension. Once appointed, the project team was able to negotiate specific projects with each contractor. A similar process was adopted for the selection of key specialist contractors, such as steel frame suppliers and mechanical and electrical services.

Benefit trading could now begin. One of the team's first successes was to negotiate the redesign of a steel frame by rearranging site cranage and other plant. This resulted in a net saving to the client of £100 000, with most benefits also being passed to the construction team.

Throughout the process the project team went out of its way to include the client, the auditor and other stakeholders in all key decisions. It was also anxious to demonstrate value for money. To do this they used their own extensive cost database and other available external data to benchmark their progress. These show that so far the local authority is succeeding in building well within the ministry cost guidelines for certain building types and making year-on-year cost reductions in relation to their own cost benchmarks – without reducing the quality of what they build.

3.3 STEP 2 – COMMITTING TO TRADING

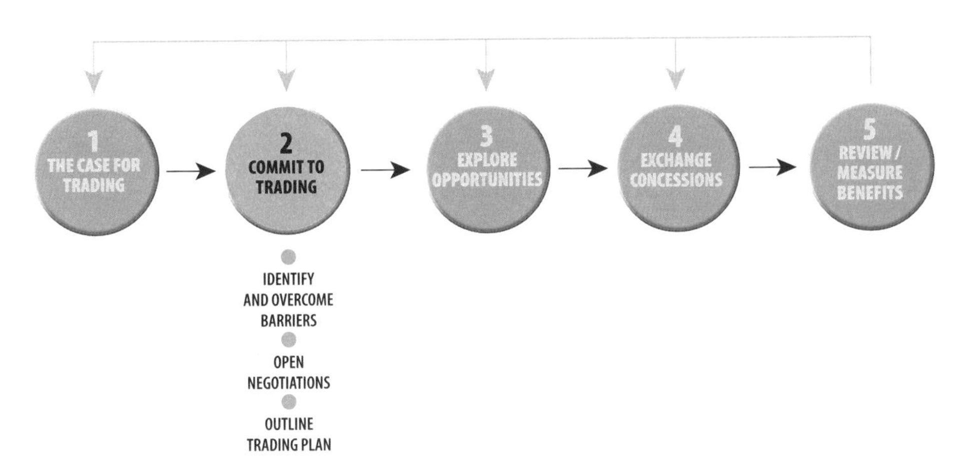

3.3.1 Recognising the barriers to trading

As well as recognising the opportunities for benefit trading, you also need to be able to present colleagues and those in authority with a sound business case for the approach. Not all will be convinced of its worth, and it may be necessary to overcome some of the more common objections levelled against it, such as:

Buyers

- "We'd like to... but I'm afraid the regulations don't allow us to do that..."
- "How do we know we are getting good value if we don't test the market competitively?"
- "Once we get into trading with a single seller he has got us over a barrel..."
- "Negotiation is open to abuse and can't be made accountable."

Sellers

- "They are such a big organisation that they'd simply eat us for breakfast..."
- "How can we be certain that they won't pinch our ideas and pass them on to someone else...?"

Each of these raises valid concerns that must be answered.

See Toolbox T4.1 *Policy and legislation*, for a summary of EC Rules for procurement

3.3.2 How buyers can overcome them

Regulations and guidance

"We'd like to... but I'm afraid the regulations don't allow us to do that..."

Many authorities have interpreted the **EC procurement Regulations** (that apply to public authorities and utilities) as a constraint that prohibits them from entering into partnering arrangements or adopting certain other procurement approaches. This should not be the case, provided that the rules are carefully followed. There is nothing in the Regulations that prevents a public authority or utility advertising its intention to enter into a collaborative arrangement for the supply of goods, services or works, provided the selection process is:

- clear as to its objectives
- fair to all tenderers
- open in its administration.

Indeed, partnering and other new construction strategies may be used as ways of demonstrating why particular suppliers provide the most *economically advantageous* offers. An agreement tendered under the EC rules may have to be retendered after a fixed period of time. However, the regulations do not prohibit an authority from developing a benefit trading approach with its suppliers, providing fairness is maintained with respect to any pre-award trading by negotiating with more than one tenderer.

Contact HM Treasury for most up to date advice (see Toolbox T5 – Useful addresses).

HM Treasury has recently reviewed and redrafted its **procurement guidance** in line with contemporary best practice and government policy on "best value". The key message is that procurers should actively seek to use selection processes, and other approaches such as partnering, that offer the purchaser best value. Benefit trading comes squarely within this category.

Most organisations, whether public or private, develop **standing orders** and other internal rules and procedures to protect them from poor practice. Procurement is frequently covered by these rules and, if the organisation has not already embraced a collaborative working culture, they may present a barrier to the use of benefit trading. Where this is the case, little progress can be made without changing the rules. The challenge is to influence change by reasoned argument and by demonstrating the benefits. A pilot study offering a "quick win" may help speed the process. We shall see in later sections that being able to demonstrate *value and accountability greatly increases the chance of success.* But first we need to address some of the concerns that lie behind many procurement procedures.

See also Section 3.6.1 for advice on benchmarking to demonstrate value.

Testing the market

"How do we know we are getting good value if we don't test the market competitively?"

As a buyer, you can – like many regular buyers – test the market, preferably by benchmarking the performance of your suppliers rather than continually seeking competitive tenders from them. After all, competitive tendering is no guarantee of lowest-priced outcomes. This kind of performance measurement is the key to demonstrating good value.

The danger of exclusive trading relationships

"Once we get into trading with a single seller he has got us over a barrel..."

A key principle of benefit trading is that each party must genuinely have something to offer the other and be willing to trade. This basis must be encouraged to prevail throughout the trading relationship – benefit trading is a way of doing business, not simply an appointment method. Both parties need to continue to find efficiencies, new offers and concessions as part of the trading deal.

See Toolbox T4.2 Delivering accountability, for further guidance.

Accountability

"Negotiation is open to abuse and can't be made accountable"

Any financial transaction that relies on judgement is open to abuse. This applies equally to competitive processes that adopt qualitative as well as quantitative criteria for making awards. The challenge, for those who must make subjective judgements, is to be clear about why decisions are made and to be able to demonstrate the basis for them.

3.3.3 How sellers can overcome them

Potential abuse of buyer's trading power

"They are such a big organisation that they'd simply eat us for breakfast..."

Again we emphasise the two-way nature of benefit trading: to do a deal both parties must genuinely be able to exchange concessions of benefit to each other. There is no point in seeking an unfair buying advantage and, by so doing, failing to make it worth the seller's time to trade. In the end this benefits no one. Buyers need sellers – the key for sellers is to be able to offer added value. If this is not possible, there is no sustainable basis for benefit trading.

Exploitation of ideas

"How can we be certain that they won't pinch our ideas and pass them on to someone else...?"

The unsatisfactory practice of cherrypicking the best ideas and passing them on to another supplier does, unfortunately, occur. While it is unlikely to occur in an established benefit trading relationship, the concern underlines two important points:

- traders should discuss and agree the business ground rules at the start of their relationship
- traders need to develop mutual trust – this can only really happen if there is genuine give and take during trading.

3.3.4 Confirming the decision to trade

By reviewing the trading context, and identifying your aims, objectives, trading opportunities and partners, you should be in a good position to decide whether to proceed with benefit trading. Of course, you need to satisfy your colleagues and, where appropriate, relevant authorities that the perceived barriers to trading can be overcome. You might, as part of your review of trading opportunities, have been able to assess the potential benefits – note that more formal evaluation of tradable issues is covered in the next, exploratory stage (Step 3) and when trading begins in earnest (Step 4). But for now, you should be happy that there is a strong case for proceeding with the approach and a good chance of a successful outcome.

3.3.5 Planning and preparation

Effective benefit trading requires thorough preparation. The most effective way to prepare is to draw up a plan. It need not be a weighty document but should ensure that:

- the objectives of trading are clearly identified and understood by the key players in the negotiating team
- the opportunities for trading are clearly identified
- the relative importance and the potential benefits of all tradable issues are evaluated
- possible outcomes (scenarios) of the trading process are considered, both up-side and down-side
- approaches are developed and evaluated, including:
  - the preferred sequence of trading
  - a statement of the roles and authority of the key players
  - an assessment of the trading positions of all parties
- arrangements are made for reviewing progress
- appropriate contingencies are made.

The plan is not a fixed document but can be adjusted as trading progresses – remember, trading involves two parties and you are unlikely to be able to anticipate all the other party's needs and potential offers to you.

3.4 STEP 3 – EXPLORING OPPORTUNITIES

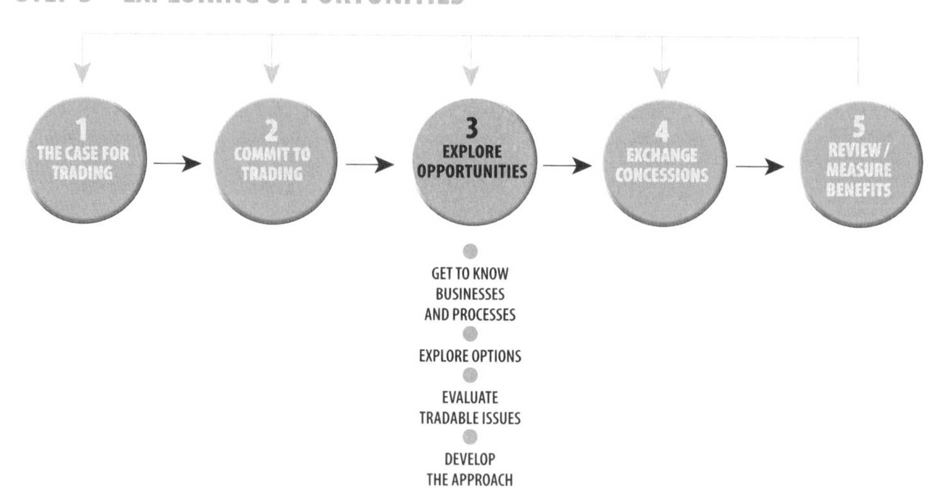

3.4.1 First moves

Depending on your circumstances, your first moves in engaging your trading partner will have begun through either:

- a direct approach (based on an existing relationship/track record)

or

- a competitive selection process.

At this point it is in the interests of both parties to ensure that the trading arrangement is soundly based and the **ground rules** are clearly established. When signing-up to a new trading relationship it is important to establish a positive basis for behaviour and, in the (unlikely) event of things going wrong, safeguard important business interests. This **trading protocol** can be especially important if there is a very unequal balance of power.

The trading protocol should be clear and simple. It is not intended to have contractual status, but is more of a charter that includes:

- an overall statement of intention
- a confidentiality agreement
- a list of off-limits topics not available for trading
- any limits on access to (particularly sensitive) information
- remuneration arrangements where the buyer orders work (including any abortive work)
- arrangements for developing trading relationships.

Precisely what is agreed under each of these headings depends on the intended trading relationship and the issues involved. If broader partnering/collaborative arrangements already exist, it may be advisable to incorporate the terms of the trading protocol into any existing partnering charter. There is nothing to stop either party revisiting and renegotiating the trading protocol in the light of experience, provided, of course, that they can both agree the changes. With the trading protocol, you and your trading partner can begin to explore trading issues and opportunities in more detail.

CASE STUDY

Alternative maintenance

A facilities manager (FM), with a £2 million annual maintenance programme, had become dissatisfied with traditional procurement and decided to develop an alternative approach offering better value for money. The new approach was based on the concept of building a team of directly employed specialist trade contractors who would work together to deliver maintenance services co-operatively. The services of main contractors were to be dispensed with altogether.

The building owner set targets for reducing maintenance costs by 15 per cent over five years. These focused on:

- overheads (tendering and administration)
- labour (efficiency savings)
- materials (reduction in waste).

Profits were deliberately ring-fenced, and rewards of increased profit margins were offered for improved performance.

How it worked

Following a thorough appraisal process, the FM appointed 12 specialist trade contractors on the basis of two-year renewable, measured-term contracts. They would each undertake work up to a value of £10 000 per project. They all had excellent track records, particularly with respect to reliability, initiative and co-operation, and each had expressed an enthusiasm to take part in, what was for all of them, a new venture. Between them, the trade contractors would be able to provide a full spectrum of construction services, from glazing to complex M&E maintenance.

The absence of a main contractor meant that, where a job involved more than one trade, the contractors had to liaise with each other and co-ordinate the work between them. So, for example, the painting contractor would be expected to negotiate his requirements for access directly with the scaffolding contractor, and so on. The specialist contractors had to overcome traditional trade demarcations and learn to develop efficient practices. A streamlined payment system was adopted by which all invoices were paid on demand by an external credit company. Random audits were conducted to ensure that the system was not being abused.

A central feature of the new arrangement was the establishment a small in-house facilities management team with overall responsibility for the maintenance and repair work. In addition to identifying work, raising orders and monitoring performance, the team's primary role was to act as a catalyst for the project by negotiating and facilitating positive interaction between the trade contractors. The team's project manager was a trained negotiator, and between them the team members aimed to develop a sufficient in-depth understanding of the repair and maintenance processes so that they would be able to initiate effective benefit trading with the specialist contractors.

The team used a number of approaches to develop trading opportunities:

- workshops – for developing new approaches and debating process improvements
- one-to-one meetings – at all levels from MD to site operatives – keeping in touch and understanding the trade contractors' business issues
- "fly-on-the-wall" observation by the facilities managers
- a structured programme of benchmarking costs and processes.

At the end of Year One, sufficient benchmarks were in place to make an objective assessment of progress. This showed that the team was well on its way to meeting the 15 per cent cost reduction target, and this was after taking account of an across-the-board award to the contractors of 1.5 per cent increased profit for the coming year.

3.4.2 Getting to know each other's business and processes/exploring options

Up to this point the basis for proceeding with benefit trading is essentially your own view of the tradable issues and opportunities. You must now begin to explore these with the other party to build up a better picture of the issues that you can trade successfully. This is best illustrated by the following diagrams.

FIG 5 TRADING NOT POSSIBLE

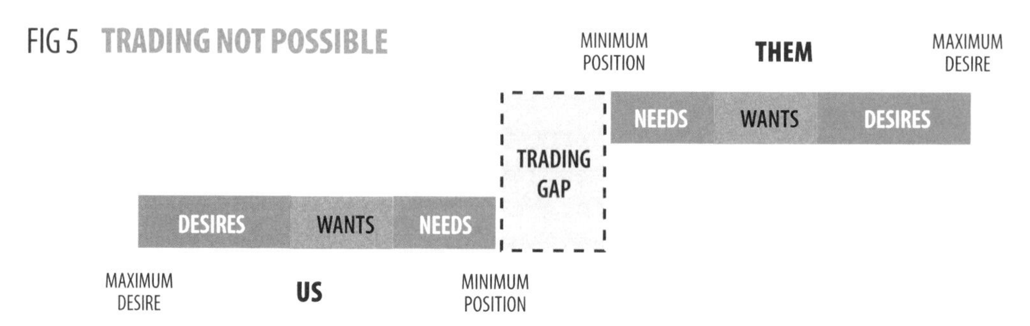

Figure 5 shows that there is no common ground between the parties and therefore no potential for trading.

FIG 6 TRADING POSSIBLE

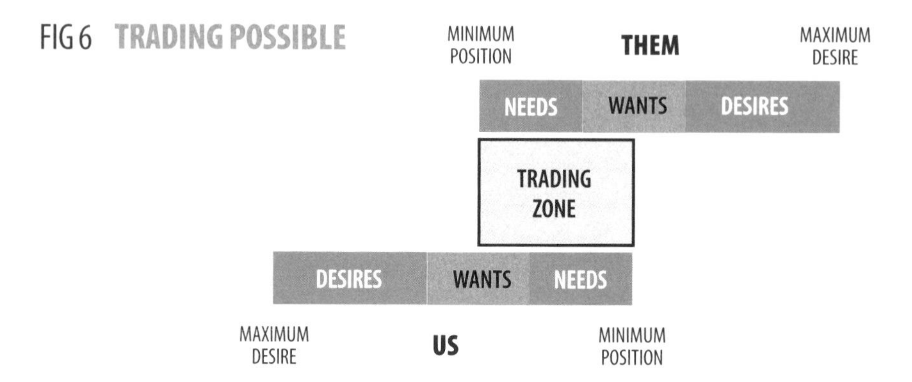

Figure 6 indicates some overlap between the party's positions and therefore suggests that trading is possible. It also suggests that it might be possible for both sides to achieve some of their wants as well as meet their needs. This point is discussed in greater detail later under the possible win-win outcomes.

Getting to know the other party's business and trying to identify more clearly the opportunities for trading requires discussions about wants and desires. You will need to demonstrate some flexibility on the tradable issues you identify, as it is unlikely that the process will run entirely according to your schedule. But your knowledge of the issues, their potential benefits and the likelihood that they are of real value is of great assistance when trying to sell your proposals later.

Working in a collaborative environment can mean that you have much more information about the other party than would normally be available in a less open relationship. It also means that more is known about your business, thereby making it harder to find new material that improves what you have to offer.

3.4.3 Evaluating potential benefits

To determine whether you have a sound basis for trading, you must attempt to evaluate the concessions you want against those that you can offer. As well as accepting that your *needs* are not negotiable, it is important that so far as possible you express your *wants/desires* in terms of what the other party is *able* to offer. While you cannot always define clearly the other party's capabilities and it will be difficult to evaluate your concessions in the other's terms, you must try to do this as objectively as possible. Try to put yourself in the your trading partner's shoes – how will your concessions be valued? What is important to you may not be so highly valued by your trading partner and vice versa. Carry out sensitivity analyses on the value of the different concessions, eg "What happens if they only value my concession x at £y 000?"

See tables 1 and 2 in Part 2 for tradable issues for key construction activities and examples of buyer's and sellers wants.

For each key concession that you want and are prepared to offer, it may help to draw up a schedule of the value to you and the potential value to the other party. Start by ranking the concessions you want and matching them against offers you may be prepared to make. This can help you clarify what you can give freely, and items for which you will need significant concessions. The value of these issues traded as concessions need not always be expressed financially and, for construction, can be expressed under five broad headings:

- performance of the delivered product
  - cost
  - time
  - quality
- appearance
- security
- convenience
- performance of the product in use
  - economy
  - durability
  - ease of use.

3.4.4 Developing the approach

Preparation and trading sequence

It is desirable to try to agree with the other party how the benefit trading process is to be structured. Matters that need to be decided include:

- need for preliminary meetings (to review information needs)
- breadth of the trading agenda (in terms of the issues to be traded)
- venue and timing of meetings
- personnel to be involved
- how communications are to be handled.

There are good reasons for deciding on a preferred sequence of trading. For example, it may be necessary to obtain the other party's agreement to one particular issue before it is possible to discuss the next; or both parties may wish to build confidence by obtaining agreement to some of the less important issues before tackling the larger ones. Whatever the reasons, it is important to agree the order in which issues are to be traded.

Who will be involved

Preparation should include consideration of who will need to be involved and what roles they will be expected to take in the proceedings. Whoever is chosen should convey confidence that they have the necessary skill, knowledge and – most important of all – the authority for the task.

It is common practice for trading to be undertaken by a small team, comprising:

- *leader*
- *summariser* – who assists the leader by reviewing agreements made and concessions outstanding as the negotiation progresses
- *recorder* – whose task is to keep track of all concessions, both suggested and agreed.

Others may be involved as appropriate. The advantage of a team approach is that individual personnel are able to observe, take notes and supply information to help the leader. Agreements reached in this way tend to be more balanced.

3.5 STEP 4 – EXCHANGING CONCESSIONS

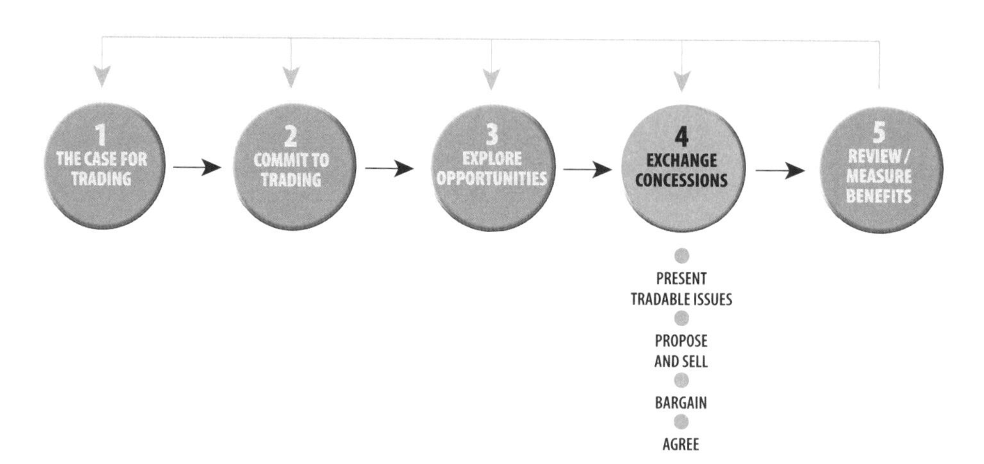

3.5.1 Getting started

Your approach

If you begin trading with a well-prepared plan, you can be confident of a favourable outcome. You will have defined your minimum needs and will know that, even if the trading does not proceed exactly to plan, you have a clear idea of what you must do in order to obtain at least some of your wants and desires.

Remember, unless there are several issues to trade in the trading zone (Figure 6) – ie it is possible for both sides to achieve at least some of their wants and, possibly, their desires – there is little point in entering trading in the first place. Remember also that win-win need not imply that each party gains the same from the negotiation.

See Section 3.4.1 for advice on Trading protocols.

Benefit trading within a collaborative working arrangement is not that different from trading within a more traditional business relationship, once both parties have decided to trade in this way. While collaborators have a mutual interest in each other's success, anyone undertaking benefit trading must work hard to identify truly valuable tradable issues and to ensure that they are rewarded fairly for their contribution.

CASE STUDY

Paint costs stripped out

A large up-market car manufacturer had set up a number of single-source business alliances with its key suppliers. At the end of a successful first year, the paint manufacturer was invited to review the company's performance and discuss future arrangements. Having exceeded the previous year's targets, the supplier anticipated using this opportunity to negotiate a price rise.

On arriving at the meeting, the paint supplier was surprised and somewhat dismayed to learn that, in the coming year, the car manufacturer was looking for an across-the-board price reduction of 7 per cent from all suppliers. And, what was more, payment would, in future, be made on the basis of the number of cars coming off the production line, rather than on the materials or components supplied.

To make such an arrangement work, the paint supplier had to be sure that his material was being applied efficiently. A visit to the paint shop confirmed the paint supplier's fears that current production methods were wasteful and poorly controlled. The paint supplier responded to the car manufacturer's proposal with a counter offer: "If you will let my company manage the paint shop, and vary the paint specification, I will guarantee to deliver equal or better quality work. And, within six months I will be prepared to accept a 7 per cent price reduction and payment for each completed car." The car manufacturer accepted this proposal subject to agreeing a number of other details. A satisfactory deal was eventually struck.

The paint supplier stuck to the bargain and by the end of the first year had met the delivery targets, achieved an improvement in the quality of finish and more than covered costs through material and production savings. Both parties were satisfied.

The trading process

For the sake of simplicity it has been assumed that benefit trading is conducted in a tidy sequential manner and that all planning and preparation is complete before formal trading begins. In practice the various trading phases may not be as clear-cut as suggested, and there may be a good deal of overlap and repetition between phases. Whatever the arrangements, it is important to bear in mind where you are in the trading process. There are four main phases:

- presenting the issues
- proposing and selling
- concessions bargaining
- making the agreement.

3.5.2 Presenting the issues

Each party should identify its starting position and the issues to be covered (the scope of the negotiations). Even in collaborative working, a certain degree of caution can be expected from both parties at this stage, as neither will probably yet want to reveal its minimum requirements from the deal. However, it is only by raising and discussing issues in an open manner that the really valuable contributions can be identified and appropriately valued. Without trust and openness you will not enjoy the full potential of benefit trading.

It is important to overcome any initial caution as quickly as possible. This is where thorough preparation will pay dividends, as you will have anticipated many of the issues raised. You can move forward by clarifying your assumptions with the other party – questions need to be positive and seek clear answers. Patient listening is required to understand and interpret what is being said.

Once you have begun to understand the other side's position, you may begin to signal where some movement may be possible. You cannot really progress until these signals have been given and recognised.

3.5.3 Proposing and selling

This is when each party tries to establish what the other is prepared to offer and whether the outcome is likely to meet their trading objectives.

See also Section 3.4.1 First moves.

Offers should always be made on a concessionary basis – "If you are prepared to do this, I may be prepared to do that...". They should also be made in the hope of moving the trading process forward, but never without the expectation of something in return. People tend not to value something for nothing and expect to have to work for their reward. This is a healthy part of even the most collaborative working relationships.

You will need to give yourself room to bargain. To get going, it is useful to start with the least important concessions – that will encourage both sides to begin trading and build the confidence and trust that will be needed for trading the major issues. Proposals do not have to be treated equally although they need to be sufficiently worthwhile to encourage movement from the other party. The validity of each offer should be carefully tested before making a counter proposal. Again, thorough preparation, particularly in the evaluation of benefits will allow a rapid and responsive review of any offers/counter-offers made.

Progress to the next phase can only be made if it is clear that the proposals made are in, or very close to, the parties' respective trading zones.

3.5.4 Concessions bargaining

Having established that a deal is possible and knowing roughly what each party has to offer, the task now is to decide how the issues should be traded. Bargaining should begin at or around the trading zone.

If the issues have been adequately defined and evaluated by both sides in their trading plans, then concessions may need to be offered in packages rather than singly. Packaging is a way of dealing with variables and presenting the issues in a form that more closely matches the other party's interests. Concessions should only be made if the other side indicates its acceptance of the conditions.

You must be prepared to retrade settled issues, if this is requested, to create movement in other areas. Unsettled issues should always be kept linked to the other party's offers/concessions so that it is clear what areas need further work from both sides before a satisfactory deal can be concluded.

In a benefit trading deal both partners expect to give and to win

3.5.5 Making the agreement

Reaching agreement is the final phase and may be achieved in a number of ways, often referred to as *closing*. The purpose of closing is to lead to agreement, and therefore its timing is critical to achieving credibility with the other side and the desired result. A *closing offer* has to be made in a manner that suggests finality and that failure to accept it will result in no deal.

Both parties must recognise and accept the proposed close. Failure to do so often results from an inability to see that the terms being offered do in fact meet their requirements, and that both sides have achieved at least part of what they set out to do.

Having agreed the deal, it is essential that all the points are recorded in detail in a way that confirms each party's understanding of precisely what has been agreed and leaves no room for ambiguity or uncertainty later. This record should be completed as soon as possible after agreement has been reached, to ensure that no issues have been overlooked or misunderstood. If the deal is a good one, all parties will be keen to implement it.

3.5.6 The people involved

People's attitudes and behaviour strongly influence the way they conduct their business. You need to understand:

- what makes a good trader
- your own strengths and weaknesses and those of the other party
- the part power plays in trading.

What makes a good trader

You are principally a problem-solver and your task is to reach a beneficial outcome, efficiently and amicably. While you focus on your own/your company's interests, you need also to have a good understanding and appreciation of the interests of the other party, and to be able to see all the issues traded in an objective light. It is not important to be liked, but it is essential that you command respect as a trader, and instil confidence that you have the authority to deliver your promises. You must be convincing in how you ask questions and raise challenges, but not in a way that might provoke adverse reactions from the other party. Trading requires perseverance, attention to detail and a flexible and enquiring mind. In short, to be a successful benefit trader, you must be:

- able to see the broad picture while remaining concerned about detail
- a good listener
- enquiring and challenging
- objective and able to walk away from an unsatisfactory deal
- able to accept and take authority
- emotionally balanced, self-controlled and even-tempered
- able to work under pressure
- flexible
- patient
- honest.

Understanding your own strengths and weaknesses

In deciding how to approach a particular assignment, you will wish to examine your own strengths and weaknesses as a negotiator. This not only assists in identifying areas for further development and training (after all, few can claim to match up to the formidable list above!), but it also gives an indication of the complementary attributes that may be required to deal effectively with other parties.

Understanding the other party's strengths and weaknesses

Section 3.4 covered the evaluation of the other party's likely concessions, and how they might value yours. Whether or not you already know the other party, it is important that you make a reasonable assessment of the *people* on the team and how they are likely to approach trading. It is important to develop a view about the best way to build their confidence and trust in you, and how you will best be able to elicit from them the information that you need.

Power and motivation

See also Section 3.4.

Power imbalances occur often in trading situations and it is important to be aware of how they arise. They can occur simply because one firm is bigger than the other, and has a purchasing power far greater than the smaller firm's potential to offer concession. However, the key to sustaining a viable trading position for the smaller party is to ensure that its concessions are valuable and valued fairly, otherwise there is little point in seeking to trade benefits.

Even in identical trading circumstances, different people will view their opportunities and threats differently. It pays therefore to understand the principle sources of one's own power and the limits of the other party's. Negotiating power arises from the following:

- *resources* – knowledge, information, money, time, people, skill
- *laws or precedent* – legitimacy, fairness, rights
- *psychological* – reputation, desire, competition, mutual support, risk-taking, commitment.

3.6 MEASURING AND REVIEWING THE BENEFITS

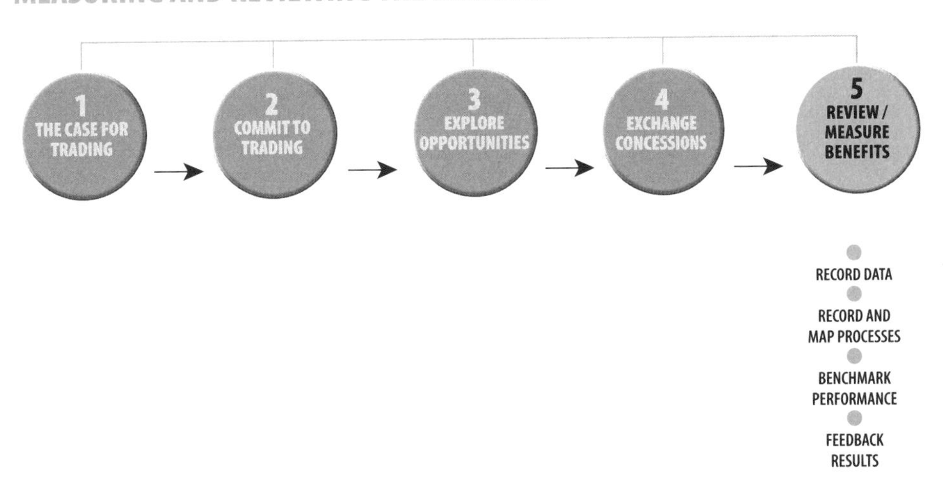

3.6.1 Benchmarking and mapping

Why measure?

There is little point in striving to improve value unless we know whether or not we are being successful. Earlier in this guide, attention was drawn to the fact that individuals or organisations have differing value requirements. The question "Am I getting good value?" cannot be answered until the values have been defined and a set of objective standards established.

Some commonly used devices, such as market testing through competitive tendering, may give an indication of relative values, but the comfort it offers may, at times, be misleading. Best practice in other industries points to a different method. In some manufacturing industries, for example, extensive databases of performance benchmarks are maintained against which suppliers are assessed and selected, without recourse to competitive tendering.

Benchmarking and process mapping are integral features of benefit trading and serve important functions, as they:

- demonstrate value being achieved
- provide the means to monitor performance
- highlight opportunities for benefit trading – including areas where performance can be improved.

Benchmarking

See also Section 3.4.3.

Participants in most collaborative alliances have the advantage of repeating what they do and are therefore in a position to measure performance against their own set of benchmarks. The headings used for evaluating benefits are equally appropriate for benchmarking:

- performance of the delivered product
- appearance
- security
- convenience
- performance of the product in use.

The function of benchmarking is to:

- measure performance against a set of defined standards
- enable the identification of opportunities for improvement by measuring the gaps between this standard and actual performance.

If you measure what you are doing, you know if you are doing it better

CASE STUDY

Activity sampling

A contractor was having problems meeting the time targets on a series of repetitive groundworks projects. The projects involved a team of ground workers and four specialist trade contractors, each responsible for one of the underground services. The contractor decided to review the on-site construction process using a short programme of activity sampling.

It was the contractor's normal practice to produce a monthly programme. This identified all the main activities on a weekly basis. These activities were agreed with the specialist trade contractors before work started.

The activity sampling survey involved independent observers in monitoring the progress of the work and recording:

- the activities in progress at each location
- the labour, materials and plant
- productive and unproductive time.

The survey revealed that, despite the attempts at detailed programming, problems were being caused by clashes in work sequences between the various specialist trade contractors. On several occasions, one or more of the subcontractors had to wait until another had finished. In all, the delays amounted to an average loss of 15 per cent of productive time per week. The reason why this state of affairs had not come to light previously became apparent when the observers questioned the gangers on the job. They saw little point in commenting on what was, after all, a regular occurrence and furthermore, there was no incentive to do the job more efficiently.

The contractor invited the gangers to assist in pre-planning their work and to alert the contractor to problems before they occurred. The gangers were naturally concerned that if productivity increased, there would be a possibility of redundancies. A solution to this potential impasse had to be negotiated. Informal discussions were held with each of the specialist trade contract teams including the on-site personnel. These quickly uncovered the key issues and enabled a deal to be agreed.

The basis of the deal was that the contractor:

- guaranteed no redundancies
- provided better on-site facilities, including improved protective clothing.

In return the specialist trade contractors agreed:

- to co-operate in the detailed planning process
- take responsibility down to ganger level for alerting the contractor to problems likely to reduce productivity.

3.6.2 Setting up benchmarks

Purpose

Before deciding to set up any benchmarks, carefully consider their intended purpose. Who is the information for? What is it intended to show? How much detail is required? What level of accuracy must be achieved? There are three simple rules. Benchmarks should be:

- measurable
- easy to apply
- relevant.

Benchmarks can be used to measure single issues or form a part of a larger performance review system. Global benchmarks are helpful in establishing the broader picture. For example, most regular building clients use simple cost m^2/quality to measure one project against another of a similar type. These global benchmarks, though crude, provide useful indicators of overall trends that can be confirmed, in greater detail if necessary, using more sensitive measures.

Guidelines

Experience in other industries indicates that it is possible to benchmark effectively using relatively few key indicators. In the oil refining industry, for example, no more than half a dozen high-level benchmarks are used to measure the performance of a complete refining plant. The lesson here is to consider carefully what are the most efficient and effective measures for your business – the wrong choice will almost certainly lead to wasted expense and frustration.

See Toolbox T3 for more information on benchmarking.

Other factors to be considered carefully in deciding what and how to benchmark should include the:

- method of obtaining data
- quality of data required, and that likely to be available
- cost of obtaining the data
- time required to analyse and communicate the results.

A limited number of targeted and appropriate measures is much better than a plethora of irrelevant ones.

3.6.3 Feedback and review of benefits

Identifying benefits

The successful application of the principles in this guide should deliver clear benefits in productivity and profitability, and your benchmarking system should identify them. There should also be other, less tangible benefits such as:

- improved business relationships
- an increased willingness to solve problems
- an interest in developing new opportunities
- greater knowledge and understanding of business needs and processes.

It is important to recognise these benefits. It is equally important to recognise when these vital cultural elements are not present, as this might indicate a lack of progress in capturing the real value of benefit trading.

Review and feedback

In any review you need to check how well you have succeeded in achieving what you set out to do initially, what you might have done differently and why.

Reviews should look at key processes, actions and outcomes. The following questions should help in designing your review:

- **Business objectives** and **trading opportunities** – are these properly aligned? Are there new priority areas for development?
- **Relationships** with trading partners – how well have these developed?
- **Other opportunities** for sharing non-core business activities – such as training, information technology etc. How well have these been exploited?
- **Training needs** – developing trading skills, cultural change, process management etc. Are these being addressed? How?
- **Benchmarking** – review of current benchmarks. Are they still relevant, up to date and achievable? Do they provide the right level of incentive? Where are the big gaps between performance and achievement?
- **Communications** – are all the key stakeholders being kept up to date and informed of important events and changes? Do they remain "bought-in" to the process?
- **The review process** – to take account of the important lessons learned coming from the detailed reviews. Are the right issues being looked at? Is the feedback getting to the points where it is required?

Types of review

Reviews can take many forms, depending on the nature of the process, the resources available, and when the review is needed. At the very least, you should look broadly at your role in benefit trading, and confirm that your original intentions behind benefit trading remain valid and are being addressed. This broad look can be carried out during benefit trading (for example, following each of the five key steps) or afterwards. It need not be a formal exercise but it would be wise to obtain the views of some of the key stakeholders on how well the process is developing, from their point of view.

A more detailed review should involve regular one-to-one interviews/discussions with individual stakeholders, and it may be useful also to conclude with a more formal meeting or workshop with all key stakeholders. This can provide a very effective way of ensuring that they obtain timely feedback from the review. Where a large supplier base is involved, feedback sessions can usefully be combined with social events, which provide excellent opportunities for networking and developing working relationships away from the normal workplace.

Detailed reviews should be held at least every six months. You may find that you need to review the process more frequently when you first start to use benefit trading. Reviews may be linked to the completion of key project stages, but it will be important to look at what has happened on each of the five key steps in benefit trading. The outcome of the review should be communicated to the stakeholders and include:

- brief summary report
- an action plan for the period to the next review.

See also Section 3.3.5 for more information on the Trading plan.

A note of caution: feedback is often forgotten in the day-to-day pressures of project life. Building in opportunities for review and feedback to the benefit trading plan will help ensure that this vital activity is an integral part of the process, and not an afterthought.

THE TOOLBOX

The reader's attention is drawn to the training material and checklists in the Toolbox and suggestions for further reading that may be used to amplify some of the points raised and provide specific guidance on negotiating techniques.

TOOLBOX

T1 Procurement 49

T1.1 Lessons from other industries 49

T1.2 Procurement descriptions 50

T2 Trading tools 53

T2.1 Developing the trading plan 53

T2.2 Trading goals – proforma 55

T2.3 Assessing the trading opportunities – proforma 56

T2.4 Preparing your responses – proforma 57

T2.5 Trading pointers 58

T3 Measuring 61

T3.1 Benchmarking 61

T4 Policy and legislation 65

T4 .1 EC Regulations 65

T4.2 Delivering accountability 66

T5 Reading list and useful addresses 68

TRAINING MODULES
(provided in C527)

TM Instructions for course facilitators

TM.0 An overview of benefit trading

TM.1 The case for benefit trading

TM.2 Committing to benefit trading

TM.3 Exploring trading opportunities

TM.4 Exchanging concessions

TM.5 Reviewing/measuring benefits

CASE STUDY

Nissan purchasing manager on negotiation with suppliers

"Little time is wasted in negotiation tactics as the targets are clear and usually achievable. The negotiating process explores the detailed cost structure of the supply, broken down into process, design, materials, labour and overheads. Only a very small percentage of the price involves haggling, as the majority of movement comes from improvements to process and design.

Generally, negotiation is conducted in a semi-formal manner with considerable openness. Deals struck are confirmed in writing after the event. This not only has the advantage of protecting the individual but it also ensures that there are no unanswered ethical questions.

Our buyers are trained in commercial negotiating techniques and, more importantly, gaining a very thorough understanding of the manufacturing and supply processes. We expect this knowledge to enable them to propose improvements to the production processes of the suppliers they deal with."

TOOLBOX 1 PROCUREMENT

T1.1 LESSONS FROM OTHER INDUSTRIES

Collaborative working was introduced to a number of industries during the late 1980s. Notable examples, with some relevance to construction, include the offshore oil processing and car manufacturing industries. Both manage highly technical and complex design and production processes in which the supply chain plays a major role. Both use collaborative working as a means of improving performance. Motor manufacturing, although in some ways sharing less in common with construction than offshore oil, probably has the most consistently developed approach from which construction can draw lessons in the field of supply chain management and negotiation.

CAR MANUFACTURERS AND THEIR SUPPLIERS

Car manufacturers are primarily designers and assemblers and, as such, are highly dependent on their suppliers. Increasingly manufacturers have come to rely on single-source suppliers with whom they negotiate the design and supply of the components they manufacture.

Suppliers are typically selected for their ability and track record rated against a set of key benchmarks. Appointments are made at least three months before the start of design development for a new model, thus enabling suppliers to contribute to the design process at the earliest appropriate moment. The starting position for each supplier is a target price and a component specification. Negotiation begins with the start of design development.

PROCUREMENT MANAGEMENT

Most car manufacturers have departments with clear functional responsibility for design, production, quality control and purchasing. The purchasing department has overall responsibility for procurement during the design stage, working very closely with colleagues in design, quality control and production. Negotiations concerning design developments are kept separate from those about price and supply arrangements. Negotiation is continuous, occurring throughout the development of the component. Responsibility for overall management transfers from purchasing to production as soon as design is completed.

POWER BALANCE

In many instances the balance of power is firmly on the side of the buyer. Large organisations such as car manufacturers are in a position to wield considerable clout and are aggressive in the demands they make of their suppliers. Suppliers that achieve the set targets are able to negotiate a share in the benefits achieved through design and process improvement. Responsible manufacturers take care to limit the proportion of business placed with any one supplier. (In Nissan's case, for example, this is usually no more than about one-third of the supplier's total output; after all, it has no interest in seeing its single-source suppliers go to the wall.)

POINTERS FOR CONSTRUCTION

Lessons learned from manufacturing provide useful pointers for construction:

- very great care, based on clearly defined benchmarks, is taken in the selection of single-source suppliers
- the appointment of suppliers, on the basis of a target specification and price, means that development changes have to be negotiated
- appointing suppliers very early ensures they are fully involved in design development, thus affording the greatest scope for negotiation
- buyers are able to negotiate effectively with their suppliers because they have extensive knowledge of their businesses based on a wealth of data and their own in-depth understanding of production processes
- open trading between buyer and seller, where both parties are well informed about each other's businesses, removes the opportunity (and need) for unnecessary posturing and time-wasting tactics
- an imbalance in power does not necessarily preclude win-win outcomes.

T1.2 PROCUREMENT DESCRIPTIONS

TRADITIONAL – DESIGNER-LED

The traditional designer-led approach operates on the premise that the design team is able to provide all the information necessary to tender the project, normally without the need for specialist construction advice. Where expert advice is needed prior to the appointment of the main contractor, this is usually obtained speculatively (or sometimes for a fee) from a specialist firm which is then required to tender for the work in competition with others, once the main contract has been let.

For this strategy to operate with a minimum of risk to the buyer, design information needs to be substantially complete at the time of tendering for construction services. This implies that there is little room for future change and therefore limited opportunity to apply benefit trading.

DESIGN AND BUILD

The distinctive feature of the design and build strategy is that a single organisation has responsibility for both design and construction. In many cases these organisations have well-established relationships with key members of their team and are used to working together. Unlike the traditional designer-led approach, design and build allows the contractor and other specialists to work together closely as the design evolves. It can also provide opportunities for each party to make timely contributions to the development of the project.

Where a full design service is required, the initial selection process often involves a period of negotiation to develop and agree the project requirements and overall design. This process usually continues until the detailed design is settled. In some cases the client appoints a design team to undertake a scoping study as a means of defining the brief for the design-and-build process. This obviously delays the moment at which the contractor and other specialists are able to contribute to the project.

MANAGEMENT

Management approaches require a single organisation or individual to take responsibility for the management of the project – often, but not always, including the design team. This strategy is frequently adopted for complex projects where time constraints are also critical. For this reason it is usual for key specialist trade contractors and suppliers to be appointed very early in the life of the project and take an active part in the development of the design. These specialists may also be required to take responsibility for the detailed design and installation of their particular work package.

COLLABORATIVE ARRANGEMENTS

Collaborative arrangements are now strongly supported by government, not only as a means of building on the inherent advantages of teamwork, but also because they create a positive environment in which risks can be managed and expertise shared in a mutual search for innovation and improvements in value.

Most collaborative arrangements involve:

- the buyer entering into a strategic business alliance with an individual or a limited number of key suppliers
- mutual trust and openness in dealings between team members
- commitment to continuous improvement
- benchmarking
- shared rewards
- agreed procedures for dealing with problems.

Since the principles of collaboration are well documented elsewhere, there is no need to repeat them here. However, a brief review of the key characteristics relevant to benefit trading is appropriate.

See also T5: Partnering in the Team *(Construction Industry Board, 1997); and* Partnering in the Public Sector *(European Construction Institute, 1997).*

Collaborative arrangements can take a number of forms:

- single-project partnering
- serial project/framework agreements
- long-term business alliances.

Single-project partnering is, as its title suggests, a short-term relationship, set up for the life of a single job to exploit mutual advantage for the project. While this type of partnering can enhance the benefits of teamwork and eliminate unnecessary duplication, it presents fewer opportunities for long-term improvement than other forms.

Serial projects/framework agreements are longer-term alliances offering plenty of scope for developing efficient and effective joint working practices. If these are well benchmarked it is not difficult to establish the worth of this type of arrangement. Arguments against the use of serial project alliancing suggest that it is sometimes difficult to sustain the energy required to maintain partnering.

Business alliances are not specifically project-based and are aimed at developing complementary skills and services that provide mutual benefit to the business partners. Alliances of this kind are most appropriate where the parties have a regular need for each other's services.

OPPORTUNITIES FOR BENEFIT TRADING – APPLYING DIFFERENT SELECTION METHODS

SELECTION METHOD	FEATURES	COST CERTAINTY	TIME CERTAINTY	QUALITY CERTAINTY	BUILDABILITY	RISK CONTROL	FLEXIBILITY	COMMENTS
SINGLE-STAGE SELECTION	• Tenders offer services in return for lump sum • limited opportunity to negotiate before sum agreed The use of a lump sum is intended to limit the buyer's risk	✔	✔	✔	✘	✔		Designer-led approaches allow slightly greater flexibility than design and build, but this can reduce cost certainty. Single-stage selection provides limited opportunity for thorough risk assessment of construction issues.
TWO-STAGE SELECTION	• Typically the first stage is used for short-listing • stage two usually involves negotiation, which may also include design development • most two-stage tenders convert to a lump sum after the price has been agreed	✔✔	✔✔	✔✔	✔	✔	✔	The extent of negotiation in two-stage selection varies considerably.
COLLABORATIVE SELECTION	• Initial selection may involve a two-stage process or may be a part of an ongoing relationship • allows continuous negotiation, if required	✔✔	✔✔✔	✔✔✔	✔✔	✔✔✔	✔✔	Collaborations are intended to improve value and reduce risk for the buyer through feedback and review and are the most conducive context for benefit trading.

The greater the number of ✔ the greater the opportunity for benefit trading to make a contribution.

TOOLBOX 2 TRADING TOOLS

T2.1 DEVELOPING THE TRADING PLAN

The development of a trading plan is undertaken as soon as it is recognised that there has been a positive outcome to Steps 1 and 2 and that both parties have established a willingness to commit to benefit trading. The trading plan provides the framework for the following key tasks:

- identify trading objectives
- identify trading opportunities
- review and evaluate tradable issues
- plan the approach to trading
- assess and define roles
- make arrangements for reviewing benefits and performance
- make appropriate contingency plans.

IDENTIFY TRADING OBJECTIVES AND TRADING OPPORTUNITIES

- Review your trading goals in the light of your business objectives – long and short:
  - the organisation's aspiration levels
  - personal aspiration levels
  - trading goals – the bottom line
  - trading strengths and weaknesses.
- Use the **T.2.2** proforma to help you consider your own position:
  - your essential **NEEDS**
  - your trading objectives – your **WANTS**
  - the extra things you are going for – your **DESIRES**.
- Once you have reviewed your own position and identified potential opportunities and trading partner(s), complete the **T2.2** proforma again, this time from your trading partner's viewpoint.
- Before moving on, consider the obstacles that could prevent you from developing these trading opportunities and what needs to be done to overcome them.

REVIEW AND EVALUATE TRADING OPPORTUNITIES

- Use the **T.2.3** proforma to help you evaluate the trading opportunities from your own and your trading partner's point of view.
- The process of examination will force you to consider the information you will need when it comes to trading, and whether your present assumptions will need to be clarified, either before or during the trading process:
  - what information do you need to be able to trade effectively?
  - do you have access to it or is it new?
  - does your trading partner have it?
  - can you find the information or will you have to wait until you start trading?
- In a collaborative trading relationship there is little point in concealing information from your trading partner.

PLAN THE APPROACH TO TRADING

- Consider how you would like to conduct the trading process:
  - do you need to set up preliminary meetings?
  - do issues have to be traded in a particular sequence?
  - how will the agenda be managed?
  - what decisions do you want from your trading partner?
  - what concessions will you offer?
  - how will you respond to weak and strong points?
  - who should be involved in the trading?
- Use the **T2.4** proforma to think about the awkward questions you may be asked and how you intend to deal with these and sell your concessions to maximum effect.

ROLES

- Decide who should be involved in the trading process and the roles they are to play:
  - who will lead?
  - who will summarise?
  - who will record what is said?
  - are any specialists/experts required?
  - how will authority/agreement be managed?

REVIEW THE BENEFITS AND PERFORMANCE

- Decide what mechanisms should be in place to measure the benefits and trading performance (see also Toolbox T3).

MAKE CONTINGENCY PLANS

- Not all plans succeed fully – make sure you have a system in place that allows you to manage the consequences of some part of your plan not succeeding.

T2.2 TRADING GOALS – PROFORMA

WHAT TO ACHIEVE Trading goals	NEEDS Must have?	WANTS Could do with?	DESIRES Nice to have?	POSSIBLE DISADVANTAGES Need to be aware of any down-side resulting from objectives
1	2	3	4	5

This form is to help you define **Trading Goals**

Use the columns as follows:

1 Note down all the things that you would like to achieve through trading.
2 Enter in this column the minimum (quantity/quality) that you need to make a deal.
3 Enter in this column the (quantity/quality) you would like to achieve – and realistically might.
4 Enter in this column the (quantity/quality) you would be pleased to get but do not expect.
5 Use this column to identify any potential disadvantages that your trading goals might create. These might affect your minimum trading position.

Note Trading goals are often expressed as range figures.
This means that the lowest range could lie in column 2 and the upper range in column 4. A linking line may be used to indicate this.

T2.3 ASSESSING THE TRADING OPPORTUNITIES – PROFORMA

POSITION				VALUE	
STRONG POINTS Things that IMPROVE trading position	WEAK POINTS Things that ERODE trading position	ASSUMPTIONS Things that could AFFECT trading	TRADING OPPORTUNITIES POSSIBLE tradable issues	US *LMH	THEM LMH
1	2	3	4	5	6

This form is to help you understand both trading positions **Your's and your trading partner's** Use a separate sheet for each

Use the columns as follows:

1 Note down all the things that make the trading position strong and will give leverage – ability to offer work, pay more, vary time constraints etc.
2 Consider all points that weaken the trading position – deadlines causing pressure, insufficient cashflow, insufficient resources etc.
3 Some assumptions might have to be made that will affect the trading position. A range of scenarios has to be considered. This column is particularly important when considering your trading partner's position as it will help to define the information gaps that need to be filled.
4 Identify possible tradable issues
5/6 Value the tradable issues in each other's terms. At this stage broad categories will suffice – low, medium and high.*(LMH)

T2.4 PREPARING YOUR RESPONSES – PROFORMA

TOPICS Identified in trading opportunities	POINTS What points will they/we make	RESPONSES How will we/they respond
1	**2**	**3**

This table is to help you prepare:

- responses to points that may be made **by your trading partner**
  and
- responses from your trading partner **to the points you will make.**

Both trading partners will attempt to make the most of their strong points. You need to be prepared and to have thought out responses that support your trading position.

TOPICS and POINTS refer to strengths and weaknesses – both need to be covered. You need to prepare two tables:

- your points
- your trading partner's likely points.

Use the columns as follows

1 Enter in this column the general TOPIC under which different points may be made.
2 Enter in this column the POINTS to be made under the topic.
3 Enter in this column the response to POINTS in column 2.

Note The extent of preparation here will depend on the openness of the trading partners and the amount of information they already have about each other's businesses. However well they know each other, there might well be some important issues that remain confidential and that could affect the success of trading.

T2.5 TRADING POINTERS

PRESENTING THE ISSUES

The process starts by both trading partners outlining their starting positions. The object of this part of the trading process is to exchange information and, through discussion, reach a point where the potential tradable issues begin to be defined. This requires both partners to:

- outline what they are looking for
- listen carefully to what the other says and implies
- ask positive questions
- check assumptions and fill information gaps
- justify propositions
- test commitment and priorities
- check the benefits with each other
- frequently summarise their understanding of what is being said.

Discussion should be open and positive – movement to the next stage is achieved when both partners recognise that trading is possible.

PROPOSING AND SELLING

Having explored what tradable issues might be available, the trading partners test the possible bargaining options – how the tradable issues can be linked in concession trading, where there is flexibility and where there is not. To do this, partners need to:

- propose in concessionary terms "if you will…then we can…"
- make proposals rather than arguing the case
- make and consider a number of proposals
- start low in the trading zone – allow room to manoeuvre
- examine the benefits being offered from both sides.

BARGAINING

As soon as it is clear that a deal can be achieved, the partners will start the bargaining process. Partners will know what they are hoping to achieve from their preparation and from the previous session. As trading begins, the trading partners have to:

- trade concessions
- remember that small concessions help to create movement
- expect to give something for every concession you are offered – even if not of equal value
- keep unsettled issues linked
- be prepared to bring back previously settled issues if it helps to move forward
- be creative and look for good solutions for both sides.

CLOSING AND MAKING THE AGREEMENT

Both trading partners need to recognise when one or other is offering to complete the deal. Clearly both partners have to be satisfied with the deal at that stage and be committed to making it work.

Obtaining a clear agreement is essential to any deal, especially where it involves long-term relationships. In the heat of the moment it can be easy to overlook some apparently trivial point only to find later that it has important implications for the future.

On reaching agreement:

- always summarise the points agreed at the time of the deal
- do not leave any important points to be agreed later
- clarify the detail:
  - what is included?
  - what is excluded?
  - are there any minor details to complete? If so, list them and set a deadline for completion
  - is there a time limit – what happens after it expires?
  - when does the deal take effect?
  - who will be responsible for making it happen?
  - what happens next?
  - state the timescales for the actions
- make sure that every one has the same understanding of precisely what has been agreed
- resist the temptation to get a little more out of the deal after the close
- record what has been agreed.

In cases where individuals have negotiated a deal on behalf of their respective organisations, it may be prudent to make the deal "subject to confirmation". This gives both partners time to reflect on what has been agreed and fully consider the implications. There is absolutely no point in signing up to an agreement that either party will be uncomfortable with in the long term.

After the deal:

- once agreement has been reached, both partners should implement the next moves within the agreed timescales (this builds confidence)
- know who to talk to if things aren't working out as agreed
- don't leave things until there is a problem – discuss and agree a solution
- keep building trust.

SOME OTHER POINTERS

Trading does not always progress smoothly, and you should be prepared to adjust your approach in order to move the process forward. The following paragraphs cover some of the more common ways in which trading can be disrupted and provide pointers for overcoming them.

DEADLOCK

As in any negotiation, deadlock occurs when both sides find themselves unable to move or make progress. It can be unsatisfactory and is likely to cause stress and frustration – emotions that are not conducive to successful benefit trading. While benefit trading should be fair and open, deadlock can arise as a tactic to test your strength or determination. If the deadlock is genuine, you should be prepared to abandon the deal.

Breaking a deadlock can be difficult, and usually involves the introduction of some element of change. This allows both parties to reconsider their positions (and, if necessary, to save face).

DEADLINES

Deadlines create pressure and can force decisions. It is sometimes useful to set deadlines at the start of the process, rather than introduce them part way through, as this can be misinterpreted as trying to force a deal. They should be realistic, taking account of the needs of the project. A good plan will have a timetable for trading, though as the process moves on you should review your deadlines and those of the other party to see whether they could be changed to benefit the overall process.

LOCATION

The location of the negotiations can affect your performance as a negotiator. It is important to feel comfortable and at ease with your surroundings. You need to consider carefully what is most likely to enhance your performance, and the effect location will have on the other team.

INFORMAL PROCESSES

From the key steps involved it is clear that, in practice, trading may begin long before the parties formally set about the process. This is particularly true for construction where repetition is an essential feature of design and construction development. The platform in which this takes place varies but may include:

- telephone conversations
- informal one-to-one meetings
- formal design meetings
- value management workshops
- site meetings
- continuous improvement reviews
- social events
- formal benefit trading.

You need to be aware that some or all of these occasions may be used to obtain or pass information which later forms part of the formal negotiating process. Indeed it is highly likely that much of the business is conducted in an informal way, simply because design and construction development is complex and involves contributions from a wide range of participants.

TOOLBOX 3 MEASURING

T3.1 BENCHMARKING

Part 3.6 of the guide provides an overview of benchmarking. This annex offers additional practical advice on three aspects of benchmarking:

- data sources
- approaches to data collection
- process mapping.

DATA SOURCES

There are four main sources of benchmarking data that apply to all businesses, not just construction. Information may be obtained from:

- *internal business performance data*, such as the costs of previous projects, delivery times
- *other businesses performing the same core tasks/*processes, for example timber millers, cement manufacturers
- *other businesses undertaking similar functional tasks*, such as supply management, storage
- *activities common to all businesses* including such activities as accounting, invoicing.

Internal performance data provide a convenient and potentially inexpensive resource from which to establish benchmarks. Although such benchmarks allow businesses to track their own performance, they do not readily enable comparison with other similar businesses, unless the measures can be directly related to common external benchmarks.

Data from *other companies* might be more difficult to obtain but offer a potential double advantage: they enable comparisons of market performance and, in collaborative arrangements, the opportunity to compare processes as well. Much can be learned from other organisations, particularly if they are not operating in direct competition. Benchmarks that highlight areas of relatively high performance identify advantages that can successfully be benefit traded.

Data can be obtained in a number of ways:

- privately
  - own data
  - benchmarking clubs
- public databanks
  - publications and journals
  - libraries
  - internet
  - networking
  - best practice
- Specialist benchmarking organisations
  - institutions/trade associations
  - consultancies and research organisations
  - benchmarking clearing houses.

PRIVATELY OWNED DATA

Most organisations retain data as an integral part of their business processes and records of transactions. Increasingly this information is stored and managed electronically and, if appropriately structured, selected data may be analysed to provide various key performance indicators.

It is unlikely that all the information required can be obtained in this way and special arrangements will have to be made for gathering it. Where possible these should be:

- integrated with existing key management tasks to minimise disruption and burdensome administration
- recorded close to the related events/activities to which they relate.

BENCHMARKING CLUBS

Benchmarking clubs are commonly established by businesses wishing to make best-in-class comparisons and obtain information on alternative processes. Such clubs often employ an independent consultant to undertake the data collection and analysis. This arrangement has the advantages of maintaining confidentiality between members and sharing the cost of work that small companies, in particular, might otherwise find prohibitive. Details on how to go about setting up and running a benchmarking club are given in several of the publications on benchmarking listed in the bibliography.

INFORMATION GENERALLY AVAILABLE TO THE PUBLIC

Most construction participants will be aware of information published in their trade press. They might be less familiar with information available from industry initiatives. Details of these may be obtained from the organisations referred to in the reading list/useful addresses.

SPECIALIST ORGANISATIONS

Several independent specialist consultancies and research organisations undertake benchmarking studies on a commission basis, as well as publishing data. Again, some relevant addresses are included in the reading list/useful addresses.

APPROACHES TO DATA COLLECTION

DOING THE RIGHT THING?

Before deciding to benchmark, it is essential to provide satisfactory answers to the following questions:

- what are the key issues to be benchmarked?
- where does the information come from?
- is the information already available?
- how can it be obtained?
- how accurate does the information need to be?
- what is the minimum information requirement?
- what are the measures to be used?
- can the measures be related to published/best-in-class data?
- who will take responsibility for processing the data?
- are the resources available?
- what is the cost?

- what is an appropriate frequency for updating?
- what is the best method for displaying the results?
- how will the results be communicated?

These and other questions should be carefully considered before embarking on benchmarking. Failure to think the process through at the outset can lead to wasted effort and a lack of commitment from those undertaking the work.

INFORMATION MATRICES

The following list identifies some of the more commonly adopted methods of capturing and recording data:

- information matrices
- surveys
  - self measuring
  - visits – observation
  - interviews
  - questionnaires
  - mapping.

Reference was made in the main text to the need to identify performance gaps in order to understand where performance can be improved. A convenient way of doing this is to use an information matrix to highlight differences in performance. A simple example is given below.

Example of information matrix

Site	Number of operatives	On-site storage	Just-in-time delivery	Mechanical handling	Productivity
A	5	Yes		Yes	100 per cent
B	7		Yes	No	75 per cent
C	6		Yes	Yes	90 per cent

Although this matrix does not specifically indicate the cause of the varying productivity levels, it does provide sufficient information to indicate that a closer look at sites A and B could pay dividends. Such an investigation also provides the opportunity to check whether the correct indicators are being examined. Exercises of this kind can operate at several levels, from the very broad to the highly detailed. A similar approach can be applied to process mapping.

SURVEYS

The choice of method for data gathering will depend on the nature of the activity being reviewed. In some cases a simple questionnaire will suffice, in others a visit to observe and interview is required. Whichever approach is adopted the information requirements need to be carefully structured to focus on the key issues to be examined. The information obtained might also need to highlight the differences in management styles and structures, as well as the processes themselves.

Experienced practitioners recommend piloting new benchmarks to ensure that they capture the most appropriate information in the most efficient way. This is particularly important where information is being obtained from other organisations. Has the questionnaire been tested? Are the right questions being asked? Can the responses be analysed? These and similar questions need to be satisfied.

PROCESS MAPPING

Process mapping plots individual activities over time and records what happens when they interact with others. Information recorded includes:

- a description of the activity
- location
- time
- sequence of activities, and their interrelationships
- resources (labour, plant, equipment, finance, intellectual capital).

Analysis of the process map enables planners to identify the pinch-points that cause delays or inefficiencies. Process mapping is used as a problem-solving tool and to improve process efficiency. It is particularly useful for tasks that:

- are repeated frequently
- occur less frequently but are costly
- are on the project programme's critical path.

Analysis should reveal whether the problems are the result of a failure in the process planning or more fundamentally in the design itself. As the example in the activity sampling case history in Part 3 emphasises, mapping apparently straightforward tasks can produce surprising results and bring rewarding improvements.

More advanced mapping and analysis methodologies allow comparison of process methods and the generation of new models. These and other techniques may usefully be applied in the context of benefit trading, but are beyond the scope of this guide.

TOOLBOX 4 POLICY AND LEGISLATION

T4.1 EC REGULATIONS

The following general summary is intended to provide an outline of only the main features of the EC procurement rules and should not be used as a substitute for seeking the latest HM Treasury guidance. Changing attitudes to best practice procurement mean that the rules and their interpretation are kept constantly under review.

AIMS

The EC rules apply to public authorities and utilities. Their purpose is to open up the public procurement market by encouraging:

- equal access to opportunities for tendering
- open and accountable practices
- competition as a means of obtaining value
- the use of selection criteria based on best value for money.

THRESHOLDS

The rules define the conditions and operational framework governing contracting authorities. The regulations apply to all public authorities and utilities but particular requirements come into operation when the value of a proposed contract exceeds a specified "procurement threshold" for supplies, services or works. Thresholds are updated annually and vary according to the nature of the authority and type of service being procured. The Regulations apply to both design and construction contracts where they exceed the relevant threshold.

PROCUREMENT ARRANGEMENTS

Contracts negotiated from scratch are an option in only very limited circumstances, otherwise the authority must choose between an "open" or "restricted" competitive process. The Regulations set out the requirements for advertising, including timing of notices and responses and, in the case of restricted competition, the minimum number of tenders to be invited.

Broadly the rules for tendering require that:

- tender conditions shall be discriminating
- invitations to tender must clearly state the criteria on which selection is to be awarded
- all tenderers shall be provided with the same information and shall be treated in the same manner.

Public works contracts must be awarded to the tenderer that:

- offers the lowest initial price, or
- is the most economically advantageous to the contracting authority.

Under the EC rules, each of the award criteria must relate directly to the economic advantage that the contracting authority expects to gain as a result of placing a specific contract.

Recent advice suggests that the tender evaluation process should take full account of the potential for approaches such as partnering, risk management, value management, whole-life costing etc to add value to a project and thereby create economic advantage.

FRAMEWORK AGREEMENTS

Framework agreements are permitted under the Utilities Regulations, provided that they are awarded in accordance with the rules and are not used to "hinder, limit or distort competition". Although there are no specific arrangements for framework agreements under the public sector Regulations there is "an understanding with the Commission and Member States that similar arrangements can apply".

POST-TENDER NEGOTIATION

Post-tender negotiation (PTN) may be used provided that it does not involve discrimination or distort competition. (The Central Unit on Purchasing Guidance Note 1 encourages this approach and sets out the process and protocol by which PTN should be conducted.) Hence clients operating under the directives may need to benefit trade with three or more tenderers so as to maintain fair competition. Tenderers should only be excluded from this process if they stand no chance of winning.

T4.2 DELIVERING ACCOUNTABILITY

Most organisations, particularly public bodies, have a duty to account for their actions. Benefit trading ensures that you thoroughly weigh and examine your options before reaching a decision. This process provides an excellent basis on which to deliver an accountable record of your actions. As a practitioner of benefit trading you should therefore have little difficulty in satisfying the most rigorous demands of auditors or public scrutiny, provided you follow the advice offered by this guide and apply a few simple rules.

REGULATIONS AND STANDING ORDERS

Your actions should always meet or exceed the expectations of regulations or standing orders. Before embarking on a particular course of action check regulations for compliance. If in doubt, consult an auditor or other appropriate person for advice on how to proceed.

SAFEGUARDS AGAINST CORRUPTION

Always ensure that sensitive decisions are reached using a process that is as transparent as possible:

- describe the process to be adopted
- record the criteria to be used
- include others in the process
- record decisions reached and the reasons for them.

INVOLVE OTHER COLLEAGUES

If you propose to take action that may be considered controversial, involve other colleagues or those in authority. Explain what you are proposing before taking action and seek their support. Auditors are often pleased to help, and may be able to offer practical advice on how to overcome the problem. They may even do some of the work for you!

DEMONSTRATING VALUE

Your assessment of value should always state the benchmarks and other comparative data used which aspects of these are considered important and why.

ACTIONS AND DECISIONS

Always record:
- the options considered
- the merits of each option
- the perceived advantage of the chosen option.

Being accountable requires you to leave an auditable trail that enables anyone to follow the entire path that led to a particular decision or set of decisions. If in doubt, write it down and explain your reasons.

TOOLBOX 5 REFERENCE – READING LIST AND USEFUL ADDRESSES

Negotiation

Getting to yes: Negotiating agreement without giving in 1981
Fisher, R; Ury, W; Houghton Mifflin

The negotiating edge: The behavioural approach for results and relationships 1998
Kennedy, G Nicholas; Brealey Publishing

Bargaining for results 1981
Winkler, J; Heinemann Professional Publishing Ltd

Everything is negotiable 1982
Kennedy, G; Business Books

Effective negotiating 1995
Robinson, C; Sunday Times Business Skills

Benchmarking

Benchmarking for competitive advantage 1993
Bendell, T; Boulter, L; Kelly, J; Pitman Publishing

Business process benchmarking. Finding and implementing best practices 1995
Camp, RC; ASQC

Benchmarking best practice report – construction site processes 1996
Construct IT

Benchmarking the challenge 1995
DTI Department of Trade and Industry

Selecting and finding the best practice processes in other companies for your organisation to benchmark 1994 (Feb)
Smith, D; IIR Ltd Industrial Conferences – Proven Techniques for Successful Benchmarking

Measuring up 1997
Pickrell, S; Garnett, N; Baldwin, J; BRE

Other publications from CIRIA

SP113 ***Planning to build? – a practical guide to the construction process*** 1995

SP117 ***Value by competition – a guide to competitive procurement of consultancy services for construction*** 1994

SP125 ***Control of risk – a guide to the systematic management of risk from construction*** 1996

SP129 ***Value management in construction – a client's guide*** 1996

SP138 ***Specialist trading – a review*** 1997

SP150 ***Selecting contractors by value*** 1998

Available from
Construction Industry Research and Information Association
6 Storey's Gate
London SW1 3AU
Tel: 0207 222 8891 Fax: 0207 222 1708

Construction Industry Board

Constructing success – code of practice for clients of the construction industry 1997

Partnering in the team 1997

Briefing the team 1997

Selecting consultants for the team – balancing quality and price 1997

Code of practice for the selection of subcontractors 1997

Code of practice for the selection of main contractors 1997

Available from
Book Sales Department
Thomas Telford Publishing
1 Heron Quay, London E13 4JD
Tel: 0207 987 6999 Fax: 0207 537 3631

Construction Round Table

Thinking about building

Available from
The Business Round Table Ltd
18 Devonshire Street
London W1N 1FS
Tel: 0207 636 6951 Fax: 0207 636 6952

Procurement Practice and Development, HM Treasury

Guidances

No.1 ***Essential requirements for construction procurement*** 1997

No.2 ***Value for money in construction procurement*** 1997

No.3 ***Appointment of consultants and contractors*** 1997

CUP Guidances

No.1 ***Post tender negotiation*** 1986

No.12 ***Contracts and contract management for construction works*** 1989

No.14 ***Measuring performance in purchasing*** 1989

No.19 ***PTN update*** 1989

No.33 (Revised)
Project sponsorship 1994

No.40 ***The competitive tendering process*** 1993

No.43 ***Project evaluation*** 1993

No.46 ***Quality assurance*** 1994

No.51 ***Introduction to the EC procurement rules*** 1995

No.53 ***Project management*** 1996

No.54 ***Value management*** 1996

No.56 ***Debriefing*** 1997

No.57 ***Strategic partnering in government*** 1997

No.58 ***Incentivisation*** 1997

No.60 ***Supplier appraisal*** 1997

No.61 ***Contract management*** 1997

Available from
The Public Enquiries Unit
HM Treasury
Room 89/2
Treasury Chambers
Parliament Street
London SW1P 3AG
Tel: 0207 270 4558 Fax: 0207 270 5244

European Construction Institute

Partnering in the public sector – a toolkit for the implementation of post-award, project specific partnering on construction projects, 1997

Available from
European Construction Institute
Sir Arnold Hall Building
Loughborough University
Loughborough
Leics LE11 3TU
Tel: 01509 222620 Fax: 01509 260118

Department of Environment, Transport and the Regions

Constructing the team – final report of the government/industry review of procurement and contractual arrangements in the UK construction industry, 1994 (The 'Latham report').

Available from
The Stationery Office
Publication Centre
PO Box 276
London SW8 5DT
Tel: 0207 873 9090 Fax: 0207 873 8200

Rethinking construction – report of the Construction Task Force to the Deputy Prime Minister on the scope for improving the quality and efficiency of UK construction 1998 (the 'Egan report')

Available from
Department of Environment, Transport and the Regions
Publications Sales Centre
Unit 8, Goldthorpe Industrial Estate
Goldthorpe
Rotherham S63 9BL
Tel: 01709 891318 Fax: 01709 881673

Construction Clients' Forum

Constructing improvement – the clients' pact with the industry

Available from
Construction Clients' Forum
35 Catherine Place
London SW1 6DY
Tel: 0207 931 9749 Fax: 0207 828 0266

Construction Best Practice Programme

A range of services and information

available from
CBPP
PO Box 147
Bucknalls Lane
Garston
Watford WD2 7RE
Tel: 0845 605 5556
Website: www.cbpp.org.uk